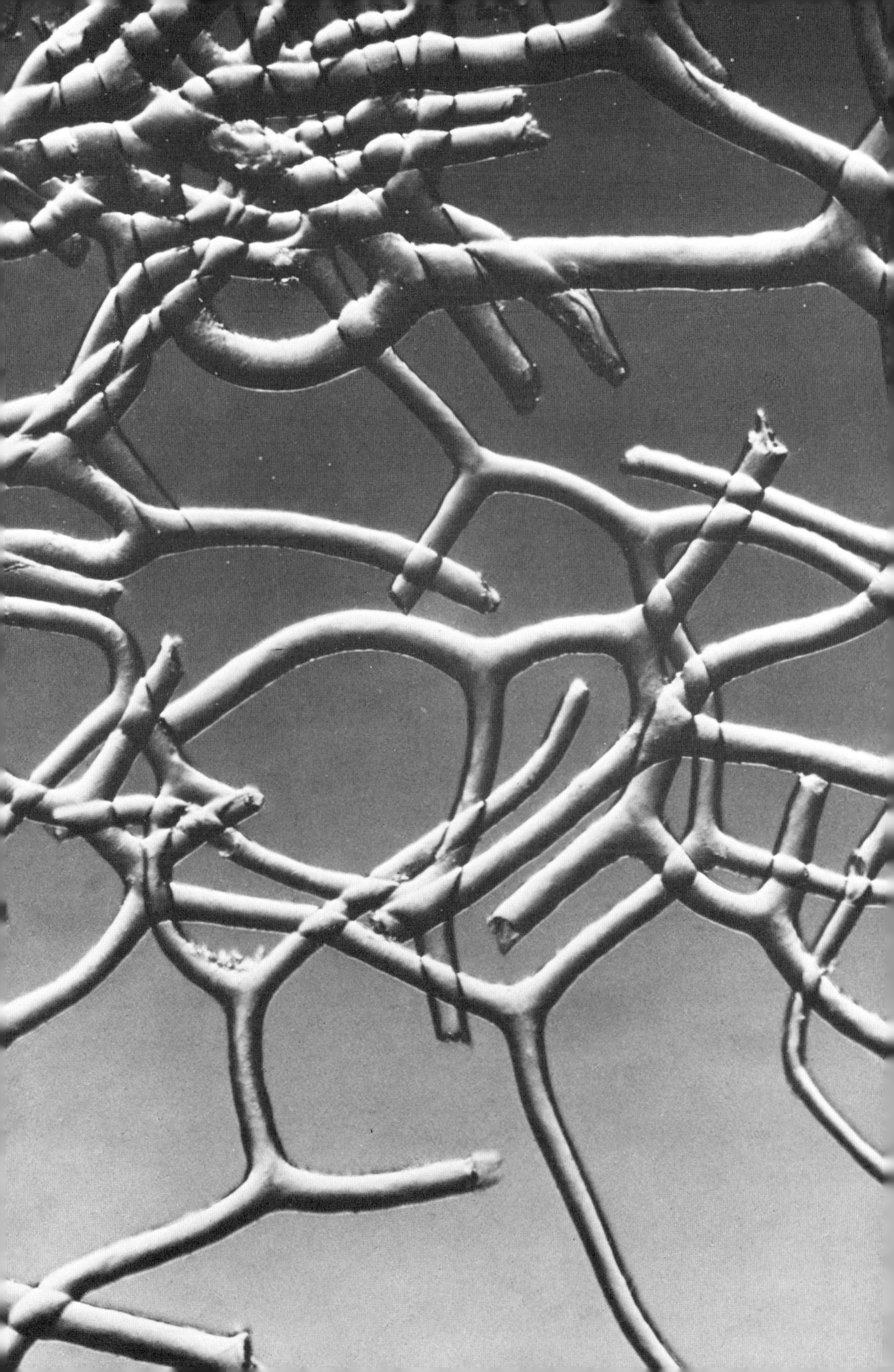

Exploring with a MICROSCOPE

This simple introduction to the microscope makes it easy to explore the world of the very small. Seymour Simon tells step-by-step how to use a microscope. He describes many fascinating things to look at and tells where to find them. He suggests some simple experiments. Many fine photographs show exactly what to look for under the lens.

Exploring with a MICROSCOPE

by SEYMOUR SIMON

Illustrated with photographs and diagrams

Random House • New York

For Robert

This title was originally catalogued by the Library of Congress as follows:

Simon, Seymour.
Exploring with a microscope. New York, Random House [1969]

82 p. illus. 24 cm. 2.95

Explains how a microscope works, gives directions for using one, describes familiar things to view that will seem unfamiliar under the microscope, and suggests some simple experiments.
Bibliography: p. [80]

1. Microscope and microscopy—Juvenile literature. [1. Microscope and microscopy] I. Title.

PZ10.S677Ex 578 70-75883
MARC

Library of Congress 70 [12] AC

Trade Ed.: ISBN: 0-394-80157-1 Lib. Ed.: ISBN: 0-394-90157-6

Cover photograph: Euglenas by Eric V. Gravé, Photo Researchers. Front endpapers: Sponge by Carl Struewe, Monkmeyer. Facing false-title: Clytia (a pond weed) by Douglas F. Lawson, Natural History Photographic Agency. Title page: Hydra catching a Daphnia from National Teaching Aids. Facing introduction: Antennae of a silkworm moth by Douglas F. Lawson, Natural History Photographic Agency. Back endpaper: Cross-section of a carrot by Carl Struewe, Monkmeyer.

The drawings on pages 10, 11, and 48 are by Ted Burwell.

Manufactured in the United States of America

Designed by Jackie Mabli

CONTENTS

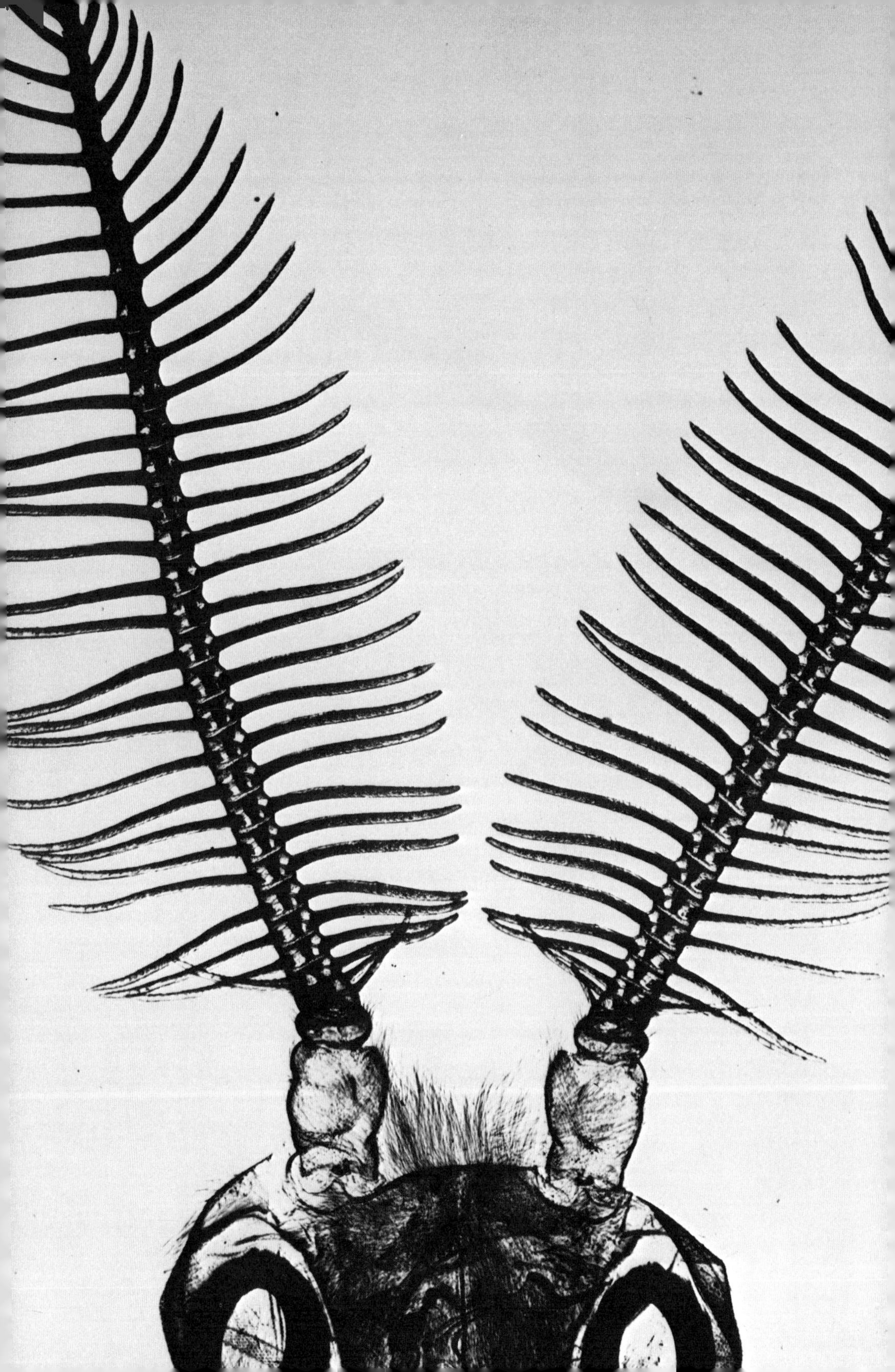

INTRODUCTION

Right now you are in the middle of a hidden world—a world too small to be seen with your unaided eyes. Your passport to this hidden world is a microscope. It makes small things look large.

With a microscope, even familiar things look unfamiliar. Under a microscope, a blade of grass becomes a stack of small green boxes, a piece of cloth becomes a weird checkerboard of lines and spaces, a drop of pond water becomes an aquarium filled with strange living things.

What can you look at under a microscope? The answer is anything and everything. From a kitchen, look at salt crystals, pepper grains, a bit of sponge, an onion skin, a slice of celery. From around your home, look at different kinds of cloth, color photographs in a magazine, dust on a window sill, a hair. From outdoors, look at a fern leaf, a bird feather, a flower, a root, a butterfly wing, or a bit of soil.

The world of the very small is full of strange and wonderful things. Even if you do not own a microscope, you can probably use one in school. Perhaps just reading and learning about the hidden world will make you want to see it for yourself. Now let's begin exploring.

THE LITTLE ANIMALS OF ANTON VAN LEEUWENHOEK

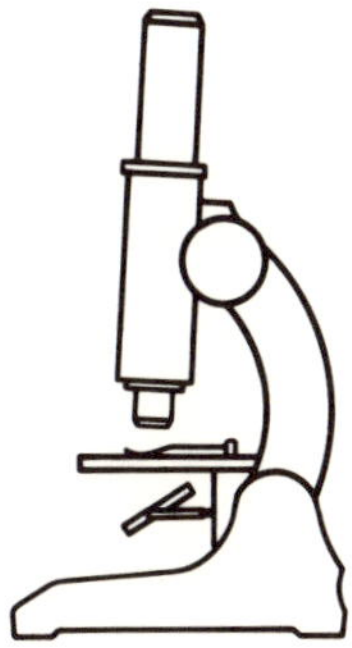

Three hundred years ago there lived in Holland a great explorer named Anton van Leeuwenhoek. Leeuwenhoek was not the usual kind of explorer. He seldom journeyed far from home. In fact, most of his trips were taken in his own home. Leeuwenhoek found a hidden world by looking through a microscope.

"In the year 1675, I discovered living creatures in rainwater, which had stood for a few days in a new pot." So begins one of the many letters that Leeuwenhoek wrote to the scientists of the Royal Society of London.

The learned men of the Royal Society read these letters with great surprise. They had never heard of such things. They found them hard to believe. They even had one of their members make a microscope to check Leeuwenhoek's findings. They marveled when he was proved right.

National Teaching Aids

Anton van Leeuwenhoek

Anton van Leeuwenhoek was born in 1632 in Delft, Holland. His family were businessmen and merchants. He left school at an early age to learn a trade in a dry goods shop, selling cloth. When Leeuwenhoek was 21 years old, he set up a store of his own in Delft. Hardly the beginnings for a scientist, you might think.

But Leeuwenhoek became interested in making lenses. A **lens** is a curved piece of glass (or any other clear material). One kind of lens makes objects look larger than they are. It is said to **magnify** objects. People may have first discovered magnifying lenses by noticing that things looked larger when seen through glass beads. A glass bead acts as a lens. By Leeuwenhoek's time, a few people knew how to carefully grind down a piece of glass into a smoothly curved lens.

Leeuwenhoek was the greatest lens maker of his time. Perhaps he made his first magnifying lens simply to examine the linens in his dry goods store. But his curiosity was such that he soon used the lens to enlarge other

Woven silk fibers as seen through a microscope. Leeuwenhoek examined many cloth fibers with his microscopes.

Carl Struewe, Monkmeyer

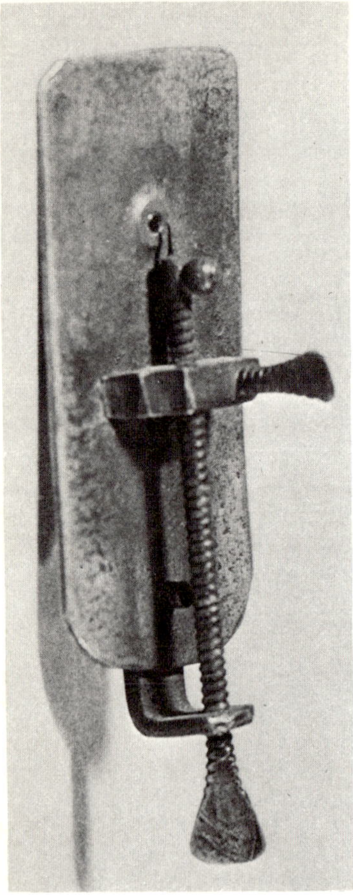

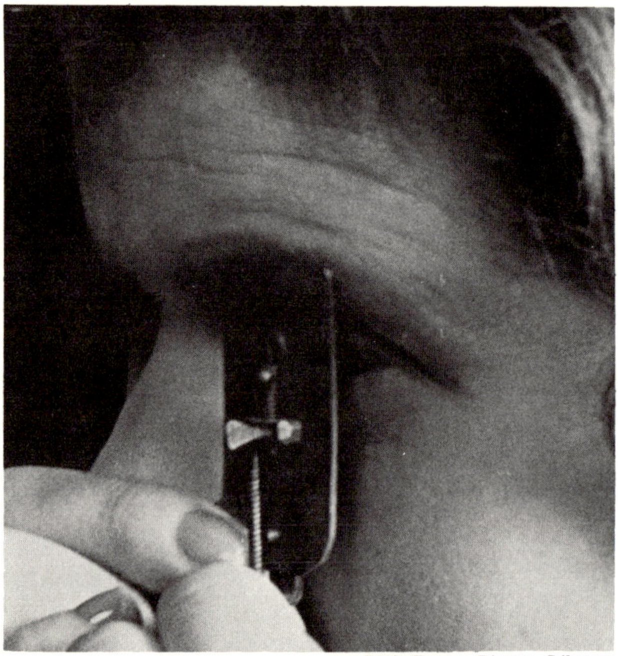

Both Ronan Picture Library

Left: a replica of Leeuwenhoek's simple one-lens microscope. Above: Leeuwenhoek's microscope in use.

objects that could not be seen in detail with his eyes alone. He looked at the head and wings of a fly, he looked at the hairs of animals, he looked at bits of dust, parts of plants, human skin, and the eye of an ox.

Soon Leeuwenhoek was spending hour after hour making more lenses and peering through them. The lenses were small but excellent. Some magnified up to 300 times. Leeuwenhoek made hundreds of simple one-lens microscopes. With them, he made discovery after discovery. He kept notes and made drawings of all that he saw.

In his notes, Leeuwenhoek wrote of tiny animals that he saw in rain water, in sea water, in vinegar, and in

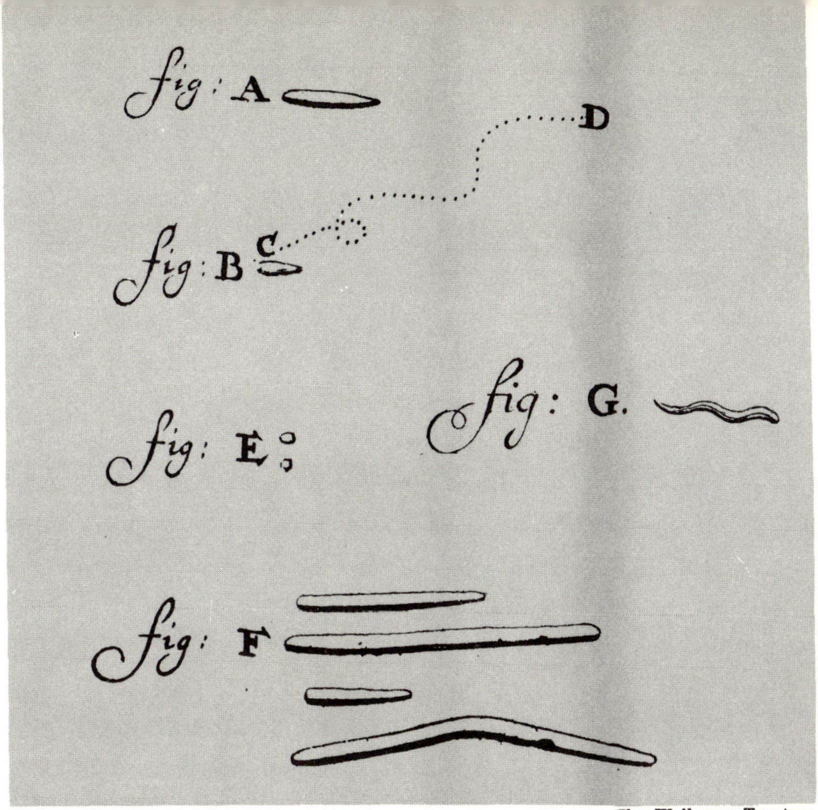

The Wellcome Trustees

One of Leeuwenhoek's drawings of the "little animals" that he saw through a microscope. These are bacteria.

mixtures of spices and water. He wrote of the way the animals moved and how they were shaped. He told of some that wriggled like tiny eels and others that moved through water "with very thin little legs."

Scraping a bit of white stuff from his teeth, Leeuwenhoek looked at it through a microscope and saw "many very little animals, very prettily moving." Some of the little animals "shot through water like a pike." Others "spun round like a top." Still others seemed to make "not the least motion." Today we call some of the "little animals" that Leeuwenhoek saw Protozoa. The ones that made "not the least motion" we call bacteria.

Leeuwenhoek was curious about everything. He examined a drop of blood and discovered red blood cells. He looked at the tail of a fish and discovered tiny blood vessels. He studied each object many times. He observed carefully and accurately.

Many important visitors came to look through Leeuwenhoek's microscopes. The King and Queen of England, the Emperor of Germany, Peter the Great of Russia, and scientists from the Royal Society all paid visits to the simple Dutch storekeeper. Other people asked to come and were put off. Leeuwenhoek was just too busy.

When Leeuwenhoek died, he and his microscopes were famous throughout the world. In the several hundred years since then, microscopes have been much improved. And nowadays, microscopes are readily available. You no longer have to grind your own lenses. Many department and hobby stores sell microscopes. With a microscope, you can rediscover the "little animals" of Anton van Leeuwenhoek. You can become an explorer in the hidden world of the very small.

GETTING TO KNOW MICROSCOPES

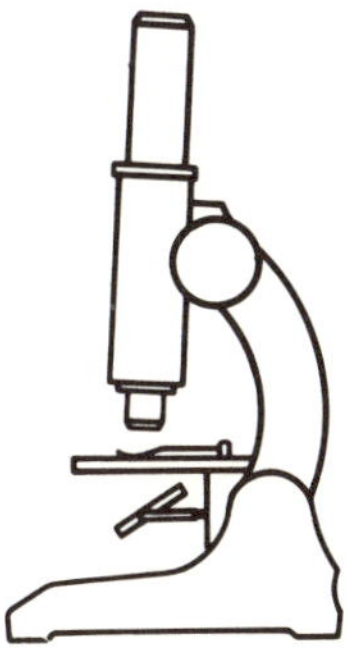

Some microscopes have a single lens. They are called **simple microscopes,** or magnifying glasses. With a magnifying glass you can examine many things around you, such as whole insects, the swirling lines in the skin of your fingertip, the parts of a flower, the crystal loveliness of a snowflake.

But to observe these things in more detail, you will need a more powerful microscope, called a **compound microscope.** Compound microscopes use at least two lenses, placed in a tube. The lens nearest your eye is called the **eyepiece.** The lens at the other end of the tube, near the object to be viewed, is called the **objective lens.** A compound microscope lets you look at things that are impossible to see even with a magnifying glass.

Both a simple microscope and a compound microscope

A magnifying glass (simple microscope)

make tiny objects appear larger. For example, a single lens that has a magnifying power of two makes an object appear twice as long and twice as wide. (This magnifying power can be written simply as 2X.) Compound microscope can make objects appear hundreds of times larger.

Suppose that you have a printed letter cut from a newspaper that you want to look at under a compound microscope. Begin by placing the microscope on a table that doesn't wobble. A microscope should be carried with

A compound microscope

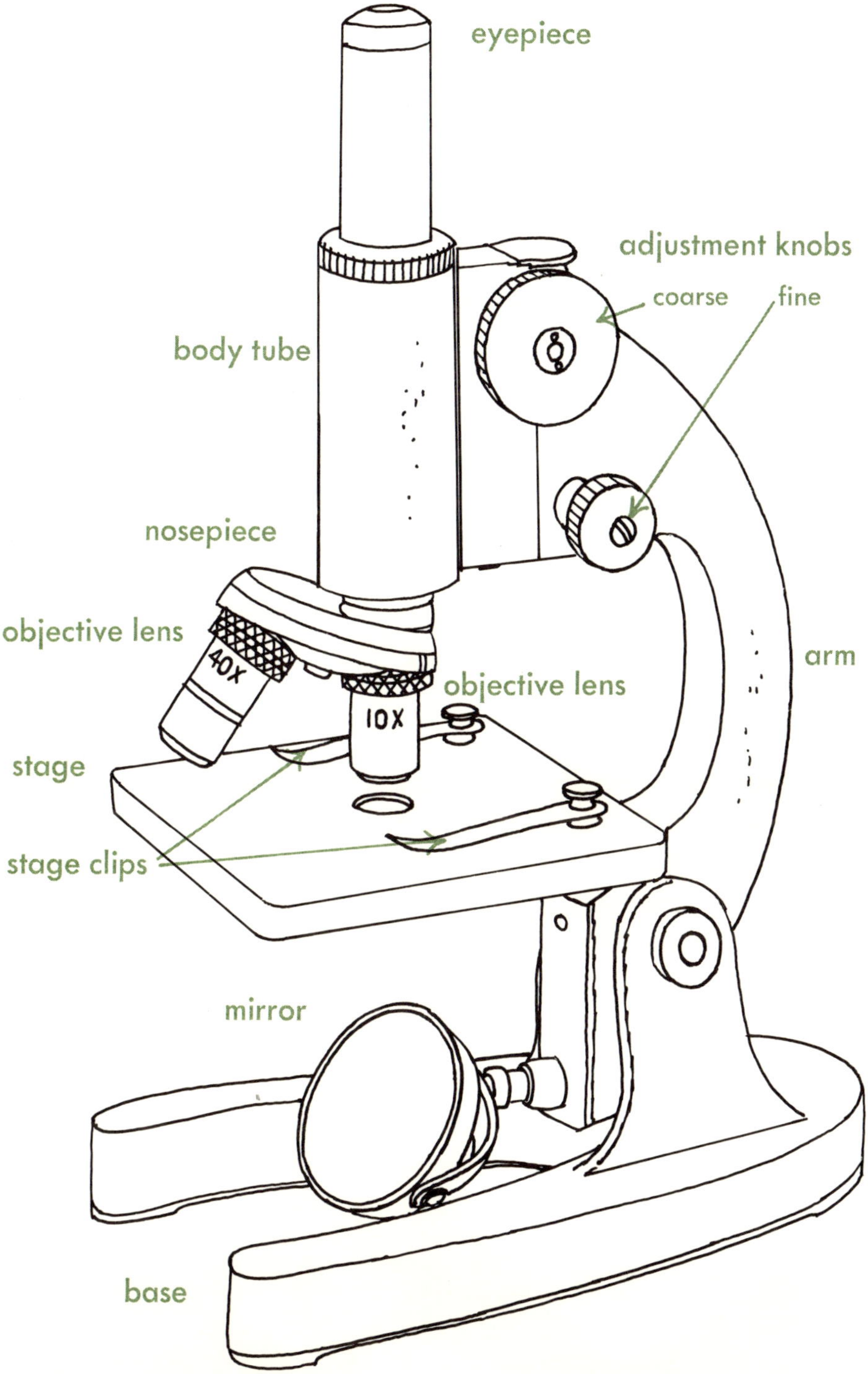

one hand under the base and the other hand holding the arm, which supports the tube. Never carry a microscope by the adjustment knobs or by the body tube.

Place your object, the printed letter, in the center of a glass **slide.** Standard microscope slides are one inch wide and three inches long. Smaller size slides that come with some inexpensive microscopes are hard to use and easy to break. Special kinds of slides that have a small cavity or well are useful for observing thicker objects.

Making a mount.

Lew Merrim, Monkmeyer

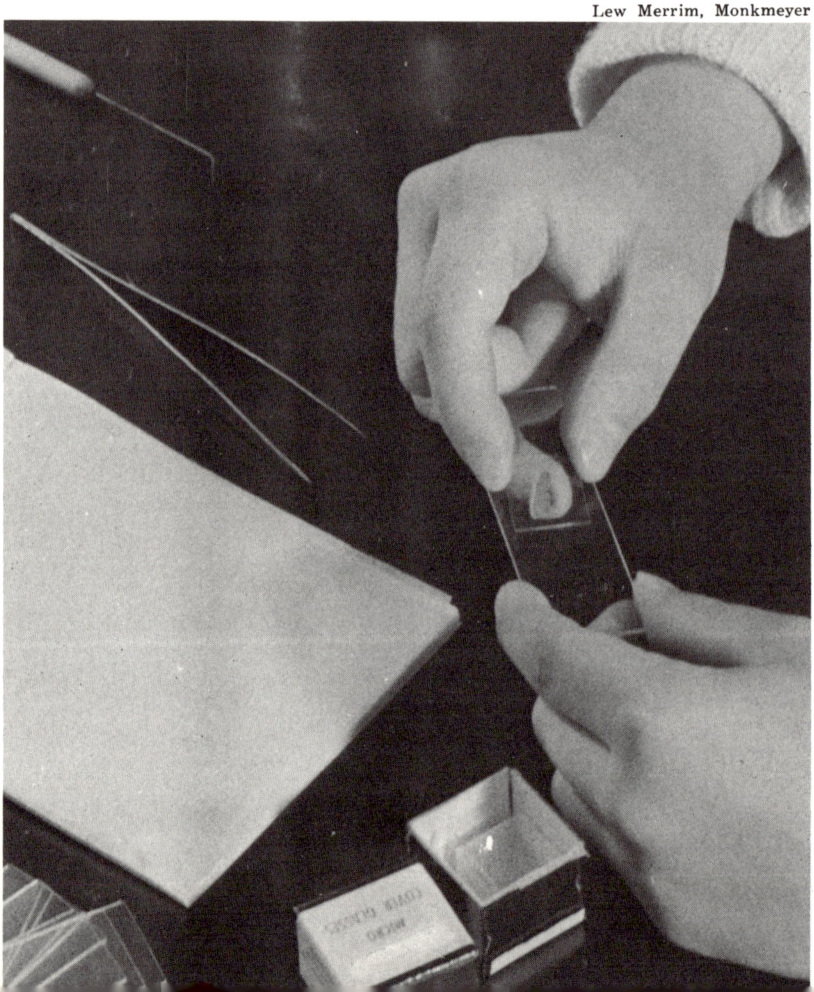

Place a **cover slip** over the printed letter on the slide. Cover slips are very thin pieces of glass or plastic that are used on top of the object, to flatten it out and keep it in place.

Slides and cover slips should be clean before you use them. Wash them with a detergent and rinse them with plain water. Then allow them to dry in the air. Always handle slides and cover slips by the edges to avoid smudging them with your fingers.

Now you have a **mount,** a slide with an object in place. Put your mount on the stage, the flat area under the tube. The slide should be exactly above the round opening in the stage. Hold the slide in place with the stage clips.

Next, you may have to choose which magnification to use. Many compound microscopes have two or three objective lenses, held in a nosepiece that can be turned. The lenses are usually marked with their magnifying power, perhaps 10X and 40X. Switch the lowest power objective lens over the stage opening.

In a compound microscope, the total magnifying power is figured by multiplying the magnifying power of the objective lens by the magnifying power of the eyepiece. For example, if the power of the objective is 10X and the power of the eyepiece is 5X, then the total magnification is 10 times 5 or 50. If you use the same 5X eyepiece but switch to a 40X objective, then the magnifying power is 5 times 40 or 200.

You may think that the bigger the magnification the

better. But this is not always so. High power lets you see only a small area under the microscope. You will see only a small part of an object greatly magnified. You may not be able to make out how this part of the object is related to the whole. It can be very confusing.

Now you can begin to **focus** the microscope—to adjust the lenses for viewing. A microscope may have one or two adjustment knobs. They move the tube up and down. At the proper distance from the object, the rays of light passing through the object come together and give you a sharp, clear image. A clear image is said to be in focus. A blurry image is out of focus. The larger knob (if the microscope has two) is called the coarse adjustment; the smaller knob is called the fine adjustment. Looking from the side of the microscope, turn the coarse adjustment to lower the tube until the objective lens almost touches the slide.

Before bringing the printed letter into focus, you will need to adjust the light. Many microscopes have a movable mirror under the stage. The mirror may be flat on one side and curved on the other. Either side can be used to reflect light up through the object and into the microscope. The curved side provides more light for high-power viewing. But the flat side often makes objects under low power appear sharper. Try both mirrors to see which is best. Work near a bright lamp or a window but **never** in direct sunlight. **Direct sunlight coming through the tube can permanently damage your eyes.** Some microscopes have a lamp instead of a mirror

Sybil Shackman, Monkmeyer

Adjusting the light.

under the stage. A lamp also gives good results.

Because the light is coming from underneath, the objects you look at through a microscope must allow some light to pass through. A piece of newspaper is not transparent, like glass, but it lets enough light through for microscope viewing.

Tilt the mirror or the lamp to give you a bright bead of light on the bottom of the objective lens. Then look through the eyepiece and adjust the mirror slightly to give a full background of light.

Now you can continue focusing. Looking through the eyepiece, bring the printed letter into view by slowly moving the tube upward with the coarse adjustment knob. At first, everything looks fuzzy. But, at a certain place, everything suddenly becomes clear—in focus Move the fine adjustment knob back and forth a little to get an even clearer image. You can see that even a slight turn of the adjustment knobs makes everything blurry again.

Always start with the tube low and move it upward. Moving the tube downward with the coarse adjustment knob while focusing may result in your breaking the slide or, even worse, in scratching the objective lens.

Under the microscope a printed letter looks surprisingly different than you might imagine. It seems to be made of dark and light smudges rather than a solid black. You can also see the coarse fibers of the newspaper. Paper fibers come mainly from the wood of evergreen trees, such as pine and spruce. Better quality paper has fibers that are close together, giving a firmer surface.

While looking through the microscope, slowly move the slide sideways. Notice that the image moves very rapidly beneath your eye. Remember that you are looking at a very small area and that even a slow movement is greatly magnified. You will also see that directions are reversed when you look through a microscope. When you move the slide to the left, the image moves to the right. When you move the slide away from you, the image moves toward you. Practice moving the slide

The letter "e" (from "Kennedy's" in the clipping at right) as seen through a microscope. The letter appears to be upside down and backwards.

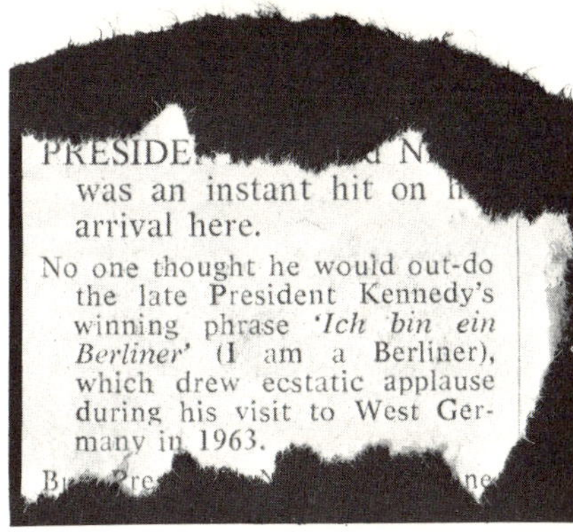

PRESIDE... N

was an instant hit on h

arrival here.

No one thought he would out-do the late President Kennedy's winning phrase *'Ich bin ein Berliner'* (I am a Berliner), which drew ecstatic applause during his visit to West Germany in 1963.

Left: the letter "e" at 16x. Below: at 60x. This means that the microscope was set at the powers of 16x and 60x when the photos were taken.

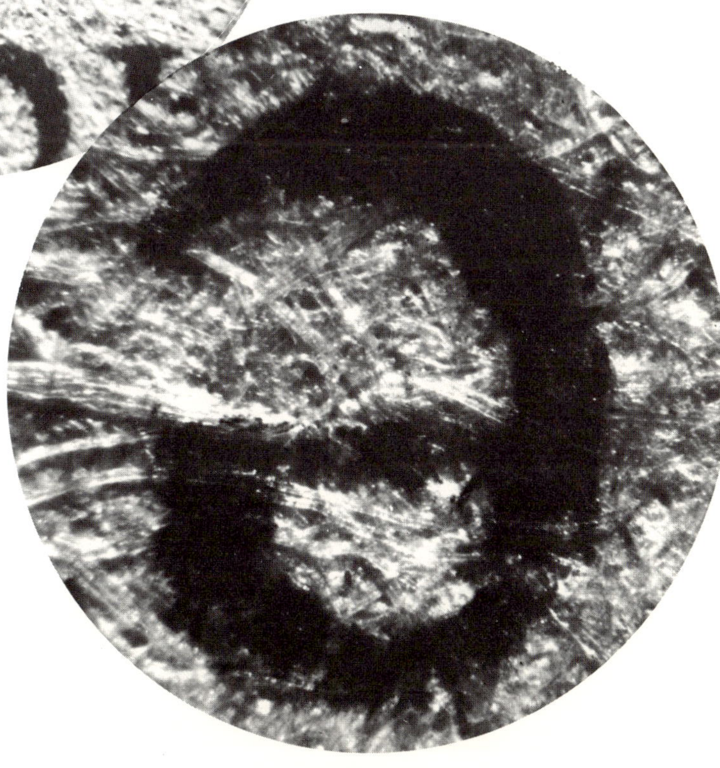

Douglas F. Lawson,
atural History Photographic Agency

NASA

Photographs in newspapers, maga-zines, and books are made up of pattern of tiny black dots. Here th area inside the black border has bee enlarged to show the dot pattern.

until you get used to the difference in direction.

Look at a bit of a newspaper photograph. To your eye, the photo looks like a series of black, gray, and white areas. When magnified, the surface of the photo becomes clusters of black dots and white spaces. Gray areas on the photo are a mixture of these black dots and white spaces. The darker the area, the closer together are the black dots. The lighter areas have just a few black dots.

Try looking at a color illustration from a newspaper or magazine. It is even more startling than a black and white one. Under the microscope, you will see bunches of black dots along with dots of several different colors, such as red, yellow, and blue. Without a microscope, your eye blends these few colors together to form the many different colors in an illustration. Look at several different color illustrations to see whether the same printed colors are always used.

For your next slide try making a wet mount. With a clean medicine dropper put a drop of milk (or diluted cream) in the center of a slide. Rest one edge of a cover slip on one side of the drop and then let the other edge down slowly. If you use too much milk, the cover will float. If the drop is too small, you will get many air bubbles. The drop should be just big enough to fill the space under the cover slip.

Under the microscope, you will see many round shapes of different sizes in a clear liquid. The round shapes

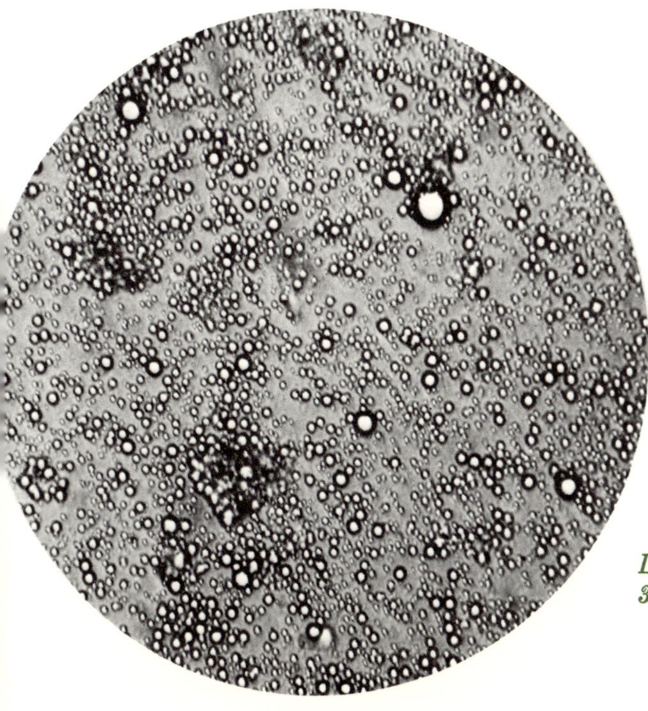

Douglas F. Lawson,
Natural History Photographic Agency

Droplets of fat in milk at 300x.

are droplets of fat. It is because of these fat droplets that a glass of milk is not transparently clear. They scatter the light coming from the outside and this makes the liquid look white.

You may see some larger round shapes that look different from the fat droplets. These are probably air bubbles that were trapped beneath the cover slip. Focus on one of the bubbles. It looks like a sparkling ring set against a darker background. Remember how air bubbles look. It is easy to confuse them with other objects for which you are searching.

Now that you know how to make slides, look around and decide what you would like to see under a microscope. Look at a dollar bill. Look at grains of salt and sugar. Look at a hair, a bit of sponge, a piece of moss. Look at everything and anything that interests you. In this book you will read about many other things to look at under your microscope.

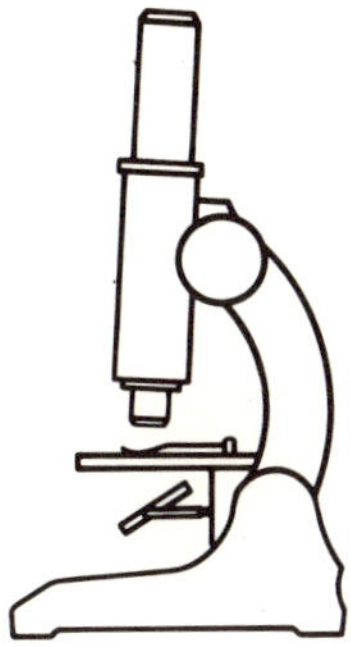

ANIMALS IN A DROP OF WATER

The water world is a great place for an explorer with a microscope. Almost any body of water contains some microscopic living things—things that can be seen only under a microscope. Ponds, lakes, streams, even puddles are home for many different kinds of life. The kinds that you are most likely to see when you first examine a drop under a microscope are called **Protozoa.**

Protozoa are tiny animals made up of a single **cell,** the smallest unit of a living thing. A single-celled protozoan is a complete animal. All by itself it does what it must to live. It takes in oxygen from the surrounding water. It takes in food—other tiny animals and plants—and releases waste products. It moves toward food and oxygen and away from harmful conditions, such as poisonous water. It reproduces more of its kind.

Sybil Shelton, Monkmeyer

Collecting Protozoa isn't difficult. Use a clean, widemouth jar. Still water is better for collecting Protozoa than swiftly flowing water. A small pond is a good place to try. Fill the jar about half-full of pond water. Collect bits of water plants, some bottom mud, decaying leaves and twigs, and some surface scum. Back at home, store the jar in a dark place away from direct sunlight.

You can also find Protozoa in your home. Water that has been in a flower vase or an aquarium for several days contains some Protozoa.

You probably will not find any Protozoa in tap water—water taken directly from a faucet. But you can use tap water to make a culture of Protozoa. A **culture** is a group of tiny living things that you are growing in order to observe. To make a culture, first boil some tap water and let it cool. Boiling the water gets rid of chlorine, a gas dissolved in the water that kills Protozoa. (Tap water will also lose its chlorine if it stands in the open for a few days.) Put some dried grass, plants, or twigs into a jar and fill it half-full of the boiled water. Let the jar stand in a dark, warm place. You should find large numbers of Protozoa in the water after a few days or a week.

Leaving your jar of pond water or your culture out in the light will produce different kinds of Protozoa. But you will not find as many different kinds, nor as large a number of Protozoa, as if you had left the jar in the dark.

Where do the Protozoa come from? Do they come from the water or the dried grass or from the air? Protozoa need water to carry on their life activities. But if the water around them dries up, they need not die. They grow a tough, protective shell, called a **cyst**, around their bodies. Thousands of these tiny cysts cover the surface of dried pond plants, leaves, grasses, and twigs. The cysts are found almost everywhere—in soil, dust, even floating in air. When conditions around the cysts become warm and moist, the Protozoa come forth once more.

Use a clean medicine dropper to place a drop of pond water or water from your own culture on a clean slide. Gently pull apart the fibers in a small piece of lens paper. (Lens paper is a special lint-free paper for cleaning lenses that is sold in camera stores.) Place the paper over the water drop. The spaces between the paper fibers help to trap and slow down the Protozoa. Without some method of slowing them down, it is difficult to examine fast-moving Protozoa under a microscope. Stand a cover slip on edge and let it down slowly over the water drop and lens paper. If you want to examine the Protozoa for more than 10 minutes, use a toothpick to spread a rim of vaseline around the edge of the cover slip. This will prevent the water drop from drying out too quickly.

Focus the microscope on the fibers of the lens paper. Move the slide around until you find some Protozoa trapped in a space between the fibers. Probably the first thing you will notice about these tiny animals is the way they move. Some kinds move rapidly and smoothly through the water. Other kinds move in spiral paths from side to side. Still others move by just flowing along very slowly. Finally, some Protozoa do not move at all, but remain attached to bits of material in the water.

One common, rapidly moving kind of protozoan is a **paramecium.** A paramecium is sometimes called a "slip-

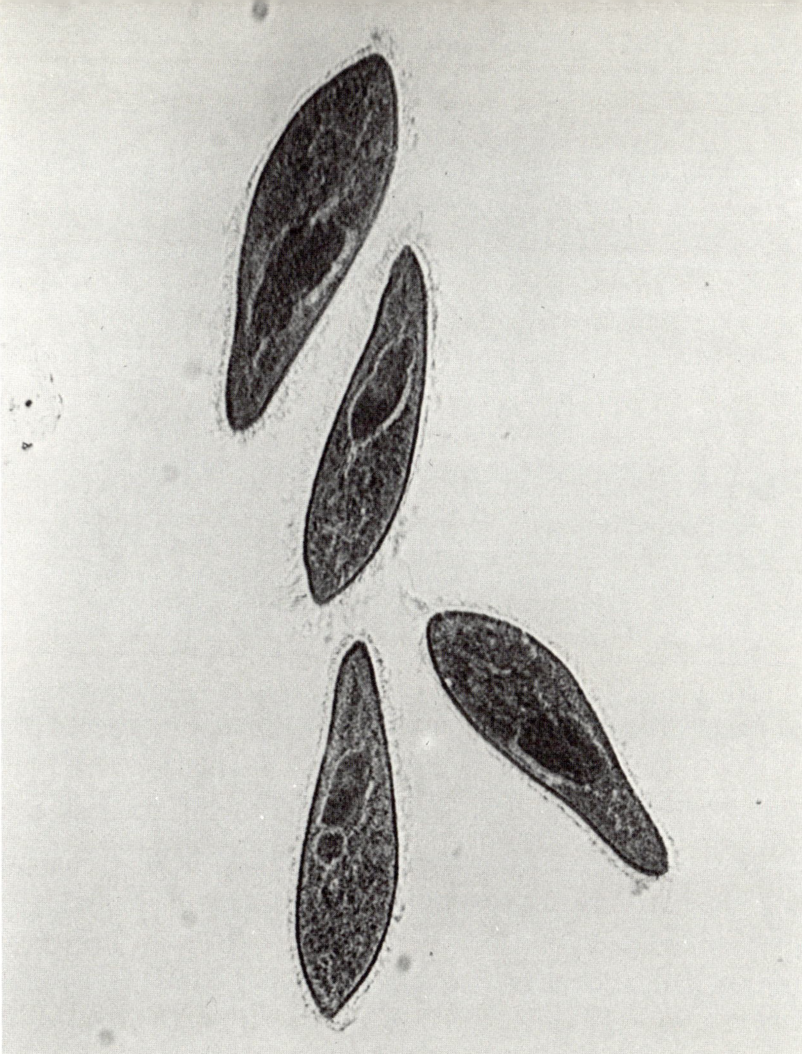

Stephen Dalton, Natural History Photographic Agency

Paramecia at 360x. The fringe around their bodies is cilia. Paramecia are actually about one-hundredth of an inch long.

per animal" because of its oval shape. The body of a paramecium is covered with short hair-like **cilia** that wave back and forth, pushing it through the water. You have to look closely to see the cilia. They are very small and beat rapidly. The cilia also push bacteria and other kinds of food into the paramecium's mouth.

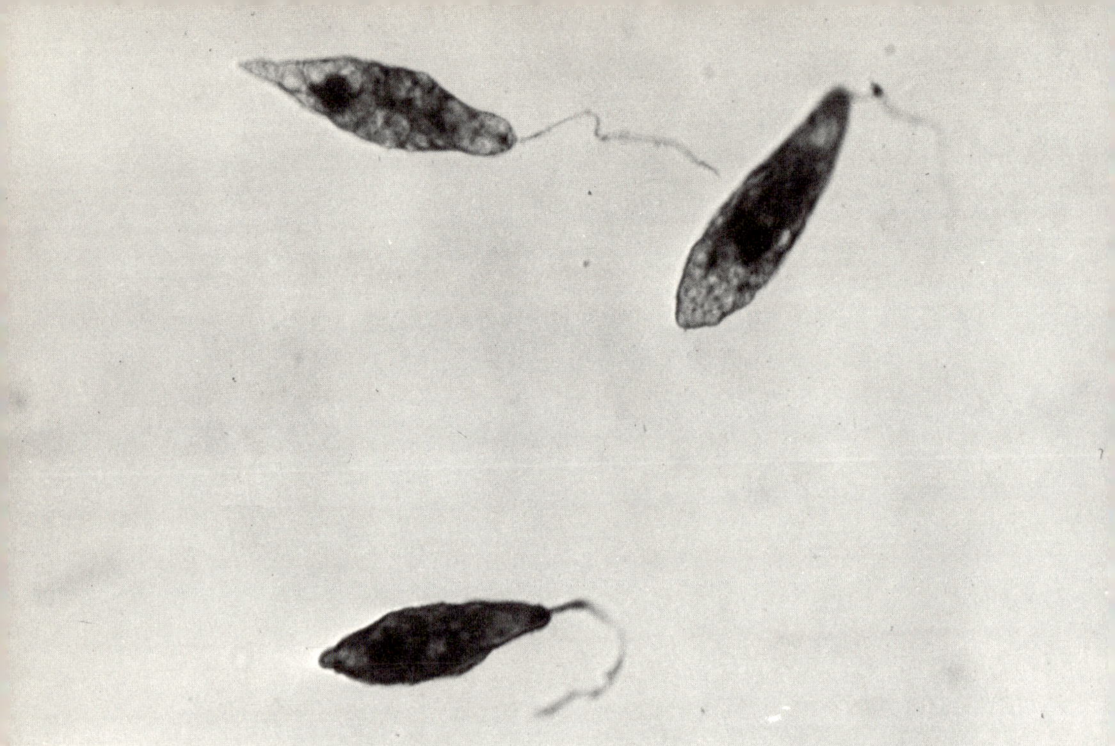

Courtesy Carolina Biological Supply Co.

Euglenas at about 200x.

A **Euglena** is a protozoan that travels in a different way. A Euglena moves by means of a long, thin tail, called a **flagellum.** A whipping motion of the tail pushes the Euglena through water, but more slowly than a paramecium. The Euglena is an odd kind of animal—if it is an animal. For one thing, it does not eat. For another, it contains bits of **chlorophyll**, a chemical found in green plants. The chemical allows a Euglena to manufacture its own food just as green plants do.

Chlorophyll manufactures food only when it is in light. Euglenas are attracted toward light. Because of this they are found mostly on the surface of water. They sense light by means of a small red dot called an eyespot. It takes a high-power microscope, at least 300X, and careful looking to see the eyespot near the front end.

Even odder than a paramecium and a Euglena is an **ameba**. An ameba has no regular shape. It moves by pushing out part of its body and flowing along after it. The pushed-out body parts are called **pseudopods,** which means "false feet." Often, several pseudopods are moving at once, but they move so slowly that you may not even notice them. In fact, it is easy to miss seeing an ameba itself. It looks like a bit of colorless jelly full of tiny bubbles and specks of dust.

When you find an ameba under your lens, watch how it uses its pseudopods not only to move but also to catch its food. One pseudopod flows around a food animal or

Amebas at 180x. These two have more pseudopods than most kinds. The arrows point to some of the many food vacuoles.

Stephen Dalton, Natural History Photographic Agency

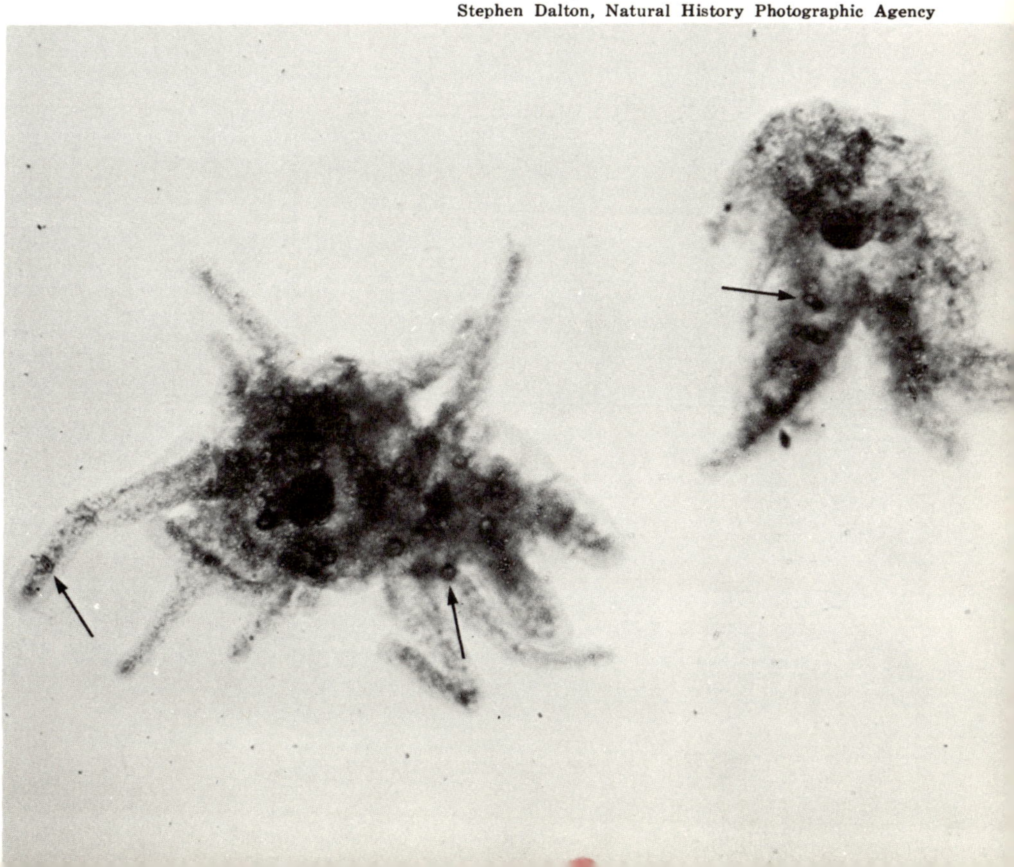

plant from one side, while another pseudopod flows around from the other side. Finally the food animal is completely surrounded by the ameba.

Trapped along with the animal is a tiny drop of water. The drop of water, called a **food vacuole,** holds the animal until it is digested. Before digestion takes place you can sometimes see the trapped animal wriggling around in the vacuole. During digestion, the ameba continues to move about and to capture more food.

Amebas and many other Protozoa reproduce by splitting into two new but identical animals. You may be lucky enough to see this happen under a microscope. The splitting takes less than an hour with an ameba. Each of the new amebas goes its separate way. Within a day they may be ready to split again. This explains why a culture of Protozoa grows so quickly. In fact, if none of the new amebas died, they would take up as much space as the whole earth within several months! Of

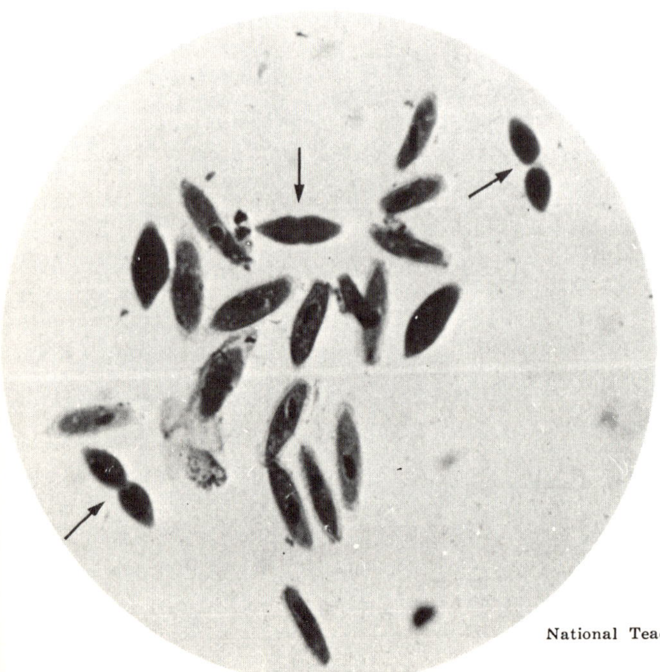

Paramecia at 10x. These paramecia have been stained with a dye to make them stand out more clearly. The arrows point to some animals that are splitting into two.

National Teaching Aids

course, amebas do die. Other animals eat them. The food supply, water conditions, and temperature often change enough to kill many amebas.

There are many kinds of Protozoa other than the ones mentioned above. You can keep records of the ones that you see by making a sketch of each new kind. Record the way it moves and behaves. Your notes should also include such information as the date and the magnification, whether there are many animals of that kind, the depth from which the pond water came, and whether the culture was kept in the light or in the dark. You can use your notes to help you identify and compare Protozoa that you see in the future.

Protozoa are not the only animals you can look at in pond water. Under a microscope, look for animals that seem to have rapidly spinning wheels instead of a head. These are **rotifers,** "wheel animals." Rotifers are many-celled animals, but they are still quite small. The spinning "wheels" are circles of hair-like cilia in motion. The cilia create water currents which push animal and plant food into the rotifer.

Some rotifers spin through the water like tiny bowling balls. Others attach themselves to plant leaves by a pointy "toe" and look like flowers. Some rotifers are brightly colored; others are colorless. Many rotifers are so transparent that you can see the inner workings of their bodies.

Rotifers are amazingly tough animals. They live not only in pond water, but in mud, rain spouts, mosses,

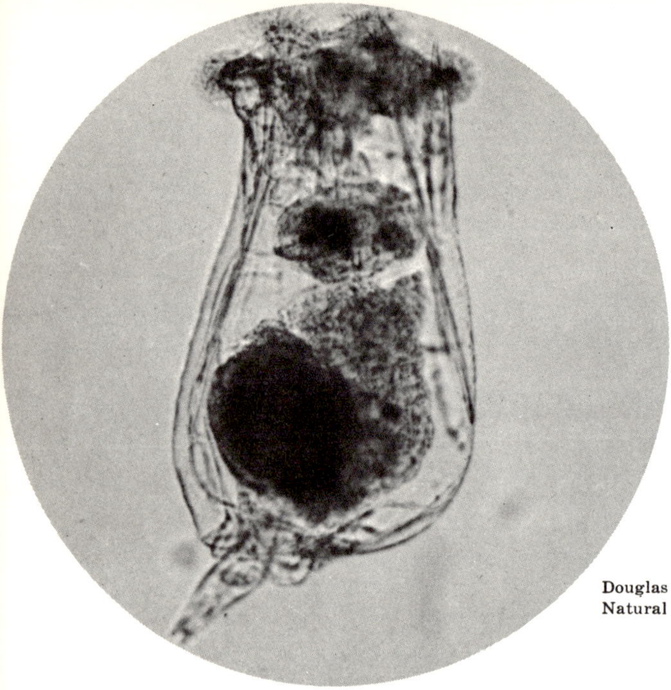

Douglas F. Lawson,
Natural History Photographic Agency

A rotifer at 150x. The whirling cilia are at top.

warm water and cold, and almost any other place where water appears even for a short time. In cold or dry conditions, rotifers form protective cysts. An aquarium that has been set up for a long time is a good place to find rotifers for microscope viewing.

Larger than a rotifer is the many-armed **hydra.** A hydra is visible without a microscope. It looks like a bit of unraveled thread hanging from the leaves of a water plant or the glass sides of an aquarium. Use a slide that has a cavity, or well, to examine a hydra under a microscope. Add a drop of water and place a clean cover slip over the cavity. Use the lowest power of the microscope.

The hydra uses its arms, which are called **tentacles,** to catch its food. When a food animal bumps into a hydra, it is shot with tiny poisoned darts that come out

A hydra, shown 16 times larger than its natural size of about half an inch long. A bud is growing out of either side.

Russ Kinne, Photo Researchers

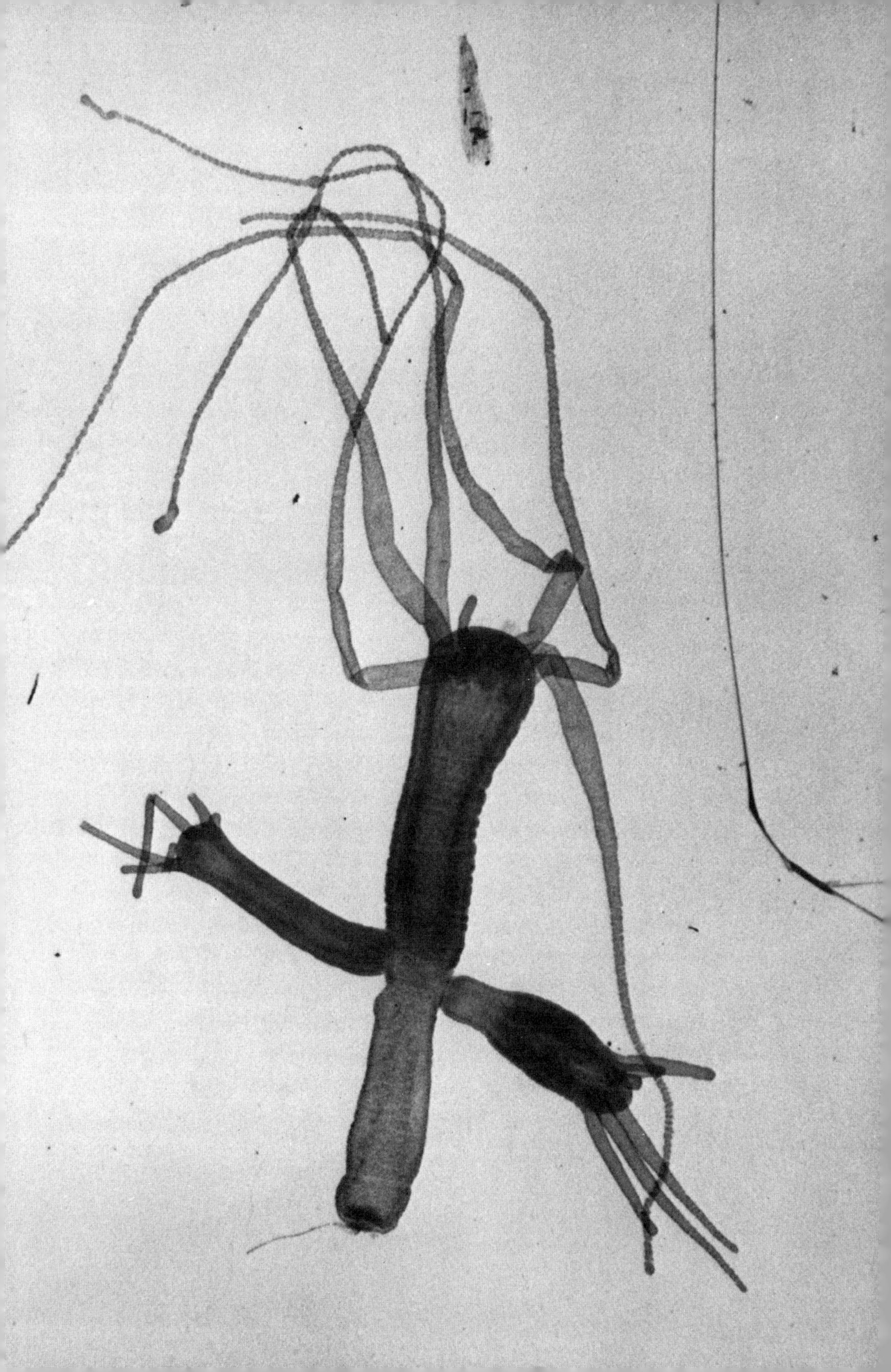

of the hydra's tentacles. Then the tentacles wrap around the animal and draw it into the hydra's mouth. Hydras are unusual in other respects, too. One way they travel is by turning somersaults. They reproduce by growing young hydras along their sides. This is called **budding.** The hydra can also regrow any lost parts. In fact, an entire hydra can grow from just a small piece!

One of the hydra's main food animals is a little hard-shelled relative of the lobster called a **Daphnia.** Daphnia are commonly sold in pet stores as food for tropical fish.

A Daphnia at about 100x. It is actually about one-tenth of an inch long.

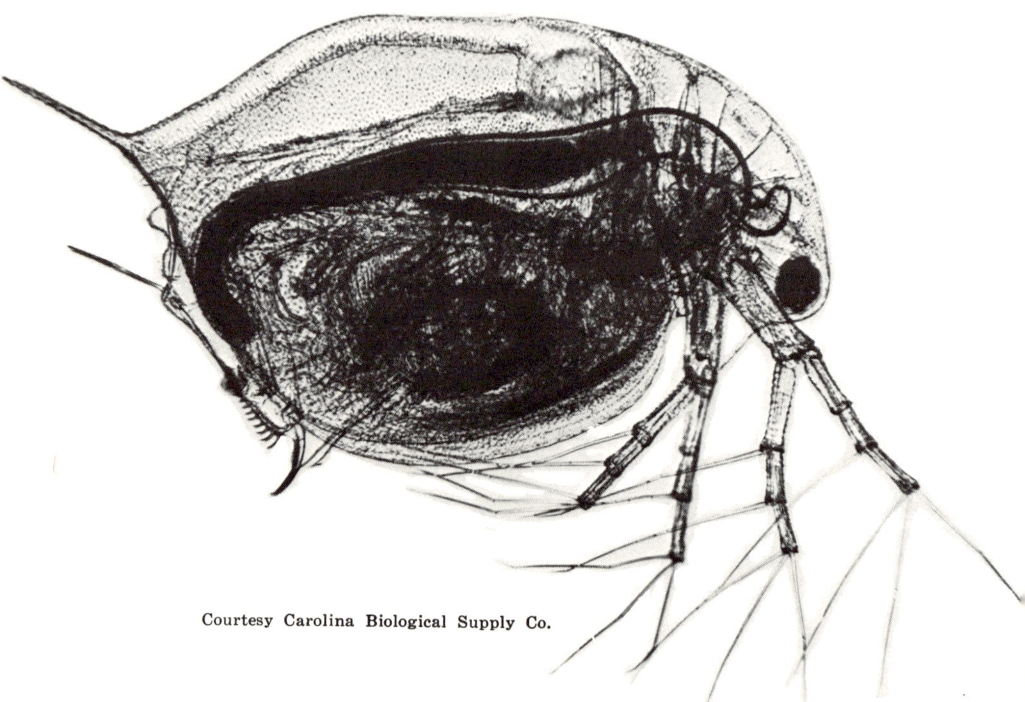

Courtesy Carolina Biological Supply Co.

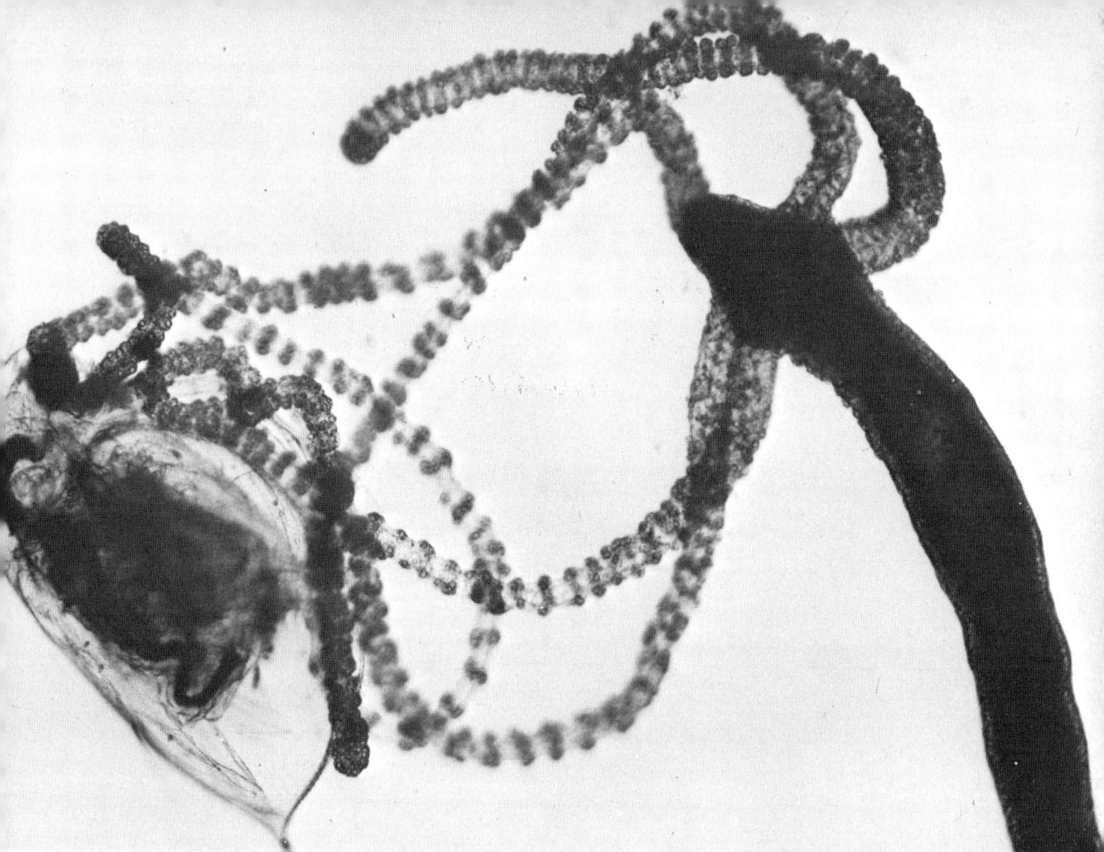

Eric V. Gravé, Photo Researchers

A hydra catching a Daphnia with its tentacles, at 50x.

Daphnia are easily visible without a microscope. They travel by hopping motions through the water. Because their hopping looks like that of fleas on land, Daphnia are sometimes called "water fleas." Just as with the hydra, use a cavity or well slide to examine a Daphnia.

Daphnia are interesting to look at under a microscope because you can see through their transparent shell and observe their organs. You can even see and count the heartbeats of a Daphnia. Try cooling or warming the slide to see whether this affects the rate of heartbeats. Put the slide with the Daphnia on it in a refrigerator or over a radiator for a minute. A **cyclops,** another small relation of the lobster, can also be used for this experiment.

Brine shrimp are related to the lobster, too. You can buy brine shrimp eggs in a pet store, hatch them in salt water, and observe the tiny shrimp under a microscope.

There are many more water animals that you can find for microscope viewing. Tiny worms, water-dwelling insects, moss animals (so-called because they look like bits of moss on leaves of water plants), snails—too many even to list. See the appendix for a list of books to help you identify these tiny water animals.

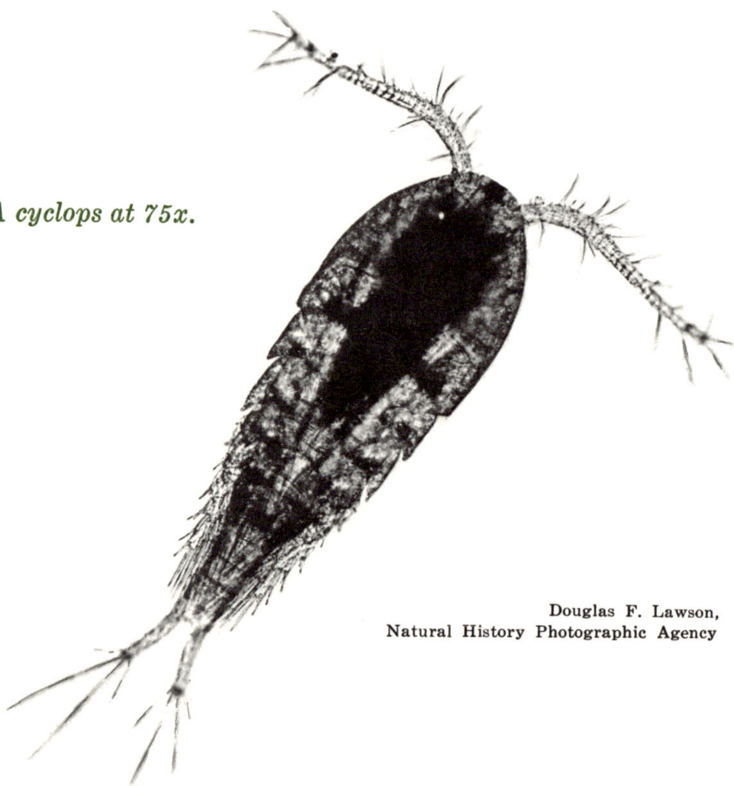

A cyclops at 75x.

Douglas F. Lawson,
Natural History Photographic Agency

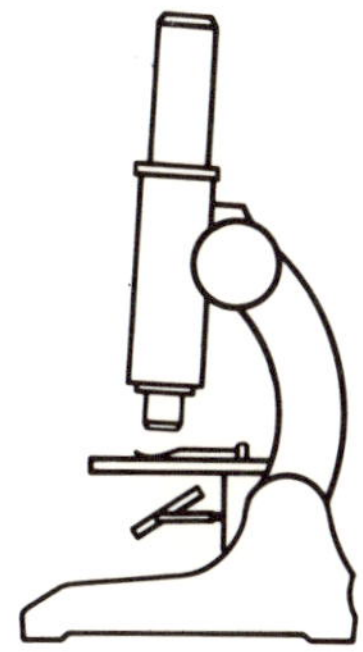

THE WORLD OF PLANTS

The giant sequoias, the maples, and pine trees all belong to the plant kingdom. Grasses, shrubs, ferns, and mosses are also plants. But not all plants are so large and easily seen. The one-celled yeasts, molds, and diatoms are small plants. You need a microscope to examine them closely.

Yeasts and molds are nongreen plants belonging to the **fungus** family. Fungi are not green because they do not contain chlorophyll. Without chlorophyll, fungi are not able to make their own food as do green plants. Fungi must obtain their food from materials around them.

Yeasts are a large group in the fungus family. The package of dry yeast or the little cake of moist yeast that you can buy at the grocery store is only one of hundreds of kinds of yeasts. This yeast is usually used in making bread. It makes the bread dough rise. The growing yeast gives off a gas that bubbles through the dough and makes it spongy.

Yeasts are also used in making beer and wine. The yeasts change sugar into alcohol. Different kinds of wines use different kinds of yeasts.

You can easily grow yeast cells to look at with a microscope. Use either a small cake of yeast or a package of dry yeast. Add a tablespoon of sugar, as food for the yeast, to a glass of lukewarm water. Place a pinch of yeast in the mixture. Stir well and keep the jar in a warm place. After a few hours, you will have many growing yeast plants to look at. If you keep the mixture for a few days, you should be able to smell alcohol in it.

Place a drop of the yeast mixture on a slide and add a drop of plain water. The water spreads out the yeast cells so that you can see them more easily. Place a cover slip over the mixture. Yeast cells are transparent and a little difficult to see. Try cutting down on the light by moving the mirrow. Some microscopes have a **diaphragm** under the stage. A diaphragm changes the size of an opening through which light passes on its way from the mirror into the objective lens. If your microscope has a diaphragm, you can adjust it to a smaller opening to cut down the light.

Look for round yeast cells and some with little bumps on them. The bumps are buds—the beginnings of new yeast cells. After a while, the bud separates from the

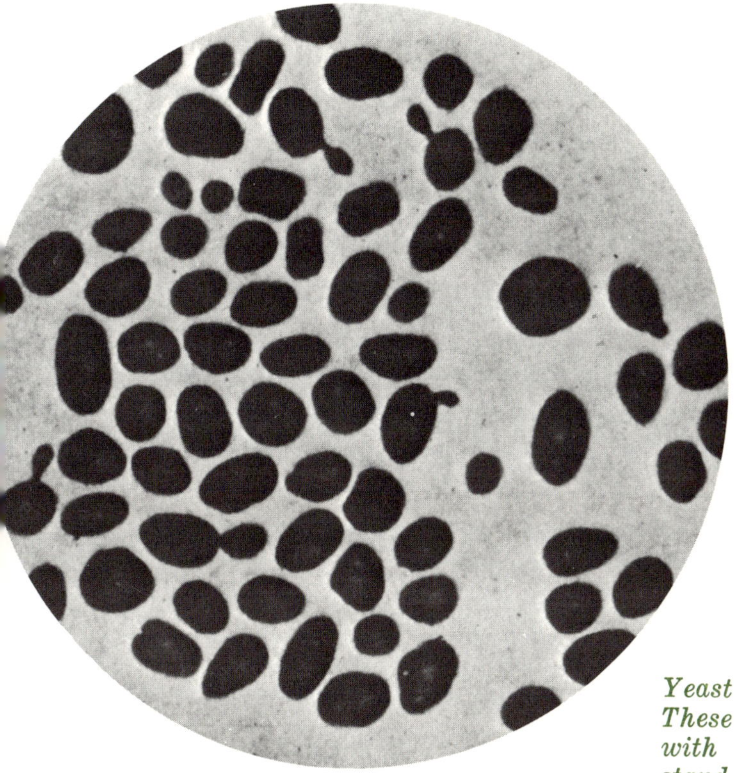

National Medical Audio-Visual Center

Yeast cells at about 600x. These cells have been stained with a dye to make them stand out more clearly. The small bumps on the cells are buds.

parent cell and is a complete yeast cell by itself. We saw in the last chapter that the hydra, a water animal, reproduces in much the same way. Yeast cells reproduce by budding as long as food is present.

A yeast cell may have more than one bud. Sometimes the new cells stick to the parent when they are fully grown and budding themselves. You will see many chains of different-sized yeast cells. The clear area in each cell is where its food is stored.

Molds make up another group of fungi. Molds grow from **spores.** Mold spores are tiny cells that are present almost everywhere. Like cysts, spores have tough walls and can withstand dry, cold surroundings for years. With the proper food, moisture, and warmth, they begin to grow very rapidly. In a short time their fuzzy growth spreads across the food.

Common foods for molds in your house are cheese, fruit, vegetables, jams, even old leather and wood. Molds may be many different colors: blue, green, pink, black, white. One of the easiest to grow for microscope viewing is the black bread mold.

Take a slice of bread and wipe it across a window sill. The dust on the sill probably contains bread-mold spores. Place a damp paper towel in the bottom of a jar, put the bread on top of the towel, and cover the jar. Keep the jar in a warm, dark place. Examine the bread each day. In a few days, you will see a fuzzy white growth with black dots scattered through it.

With a toothpick take a little bit of the mold and place it on a slide. Add a drop of rubbing alcohol or water. (Alcohol is preferable, but be careful not to use it near a flame.) Place a cover slip over the slide. Under the microscope you will see a network of branching threads. The threads act as roots, stems, and runners. Move the slide to a place where several of the threads come together. At the end of one of the threads is a small black ball, the spore case. This contains large numbers of spores. Perhaps you will find a spore case that has split, releasing its cargo of tiny spores. They

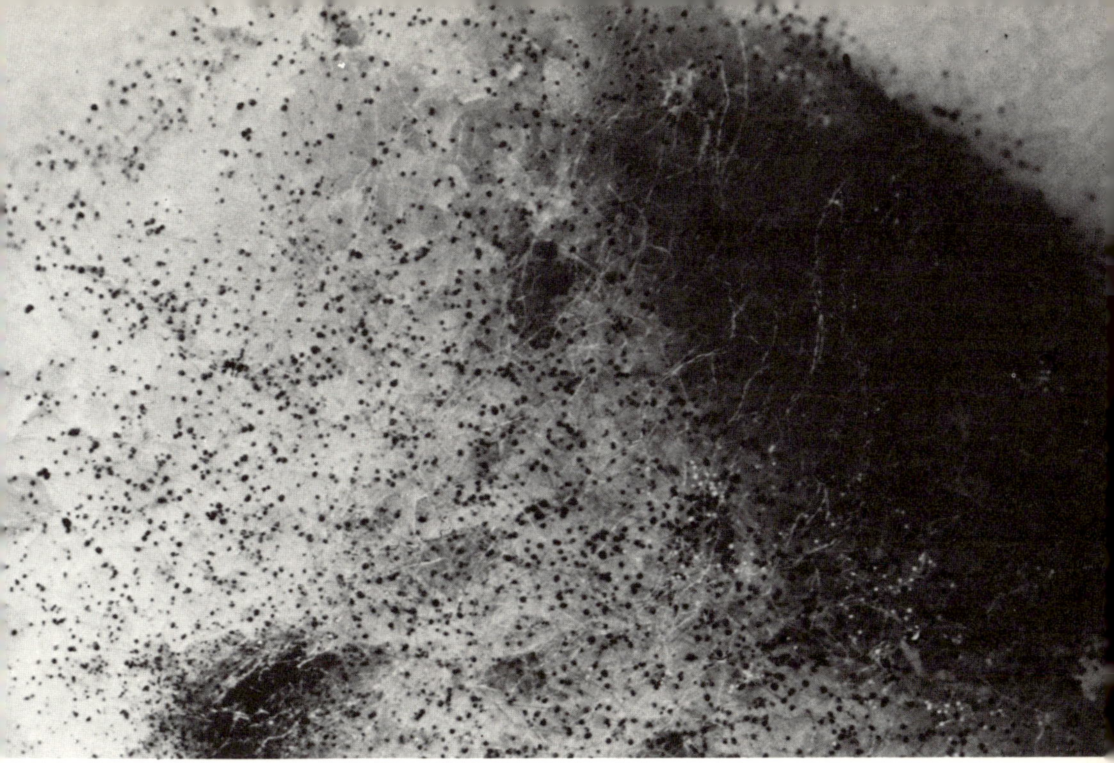

Black bread mold, enlarged below to show the spore cases. At about 10x and 100x.

Both courtesy Carolina Biological Supply Co.

look like black specks scattered around the spore case. The spores are so small that you will understand why you have never noticed them in dust.

If you keep your moldy bread in the jar for a few more days you may find a bluish-green mold also growing on it. This mold is called **Penicillium.** A medicine that kills bacteria, penicillin, is taken from this kind of mold. If you cut open an orange and transfer a bit of the Penicillium mold to it, you will get a good growth. You may see some tiny golden-brown droplets in a large growth of Penicillium. Penicillin comes from these droplets. You can examine this mold under a microscope in the same way you looked at the black bread mold.

Algae are another group of plants that you will enjoy looking at. Unlike the fungi, algae are green plants that contain chlorophyll. With sunlight, water, and the gas

Droplets in a Penicillium mold, at about 100x.

Courtesy Carolina Biological Supply Co.

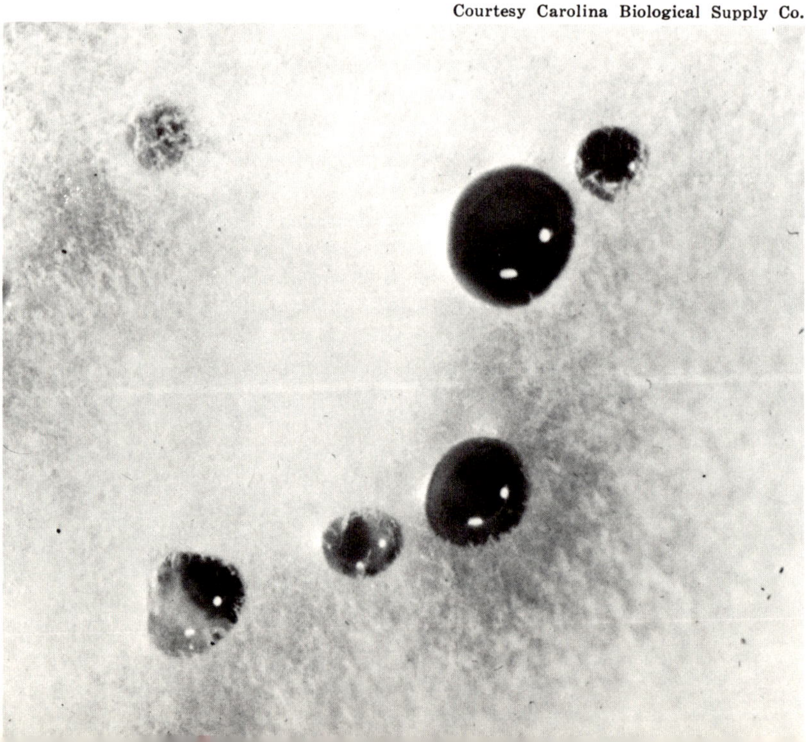

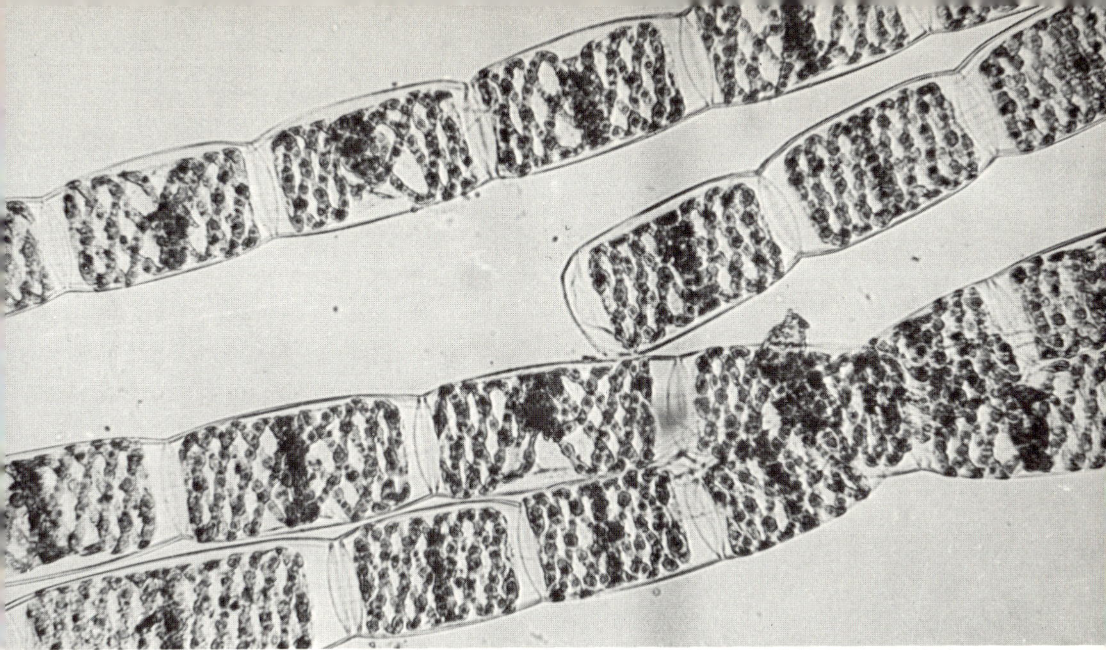

Hugh Spencer, Monkmeyer

Spirogyra at 200x.

carbon dioxide, chlorophyll-bearing plants can make their own food. Algae are common green plants, and scientists are finding out how they can be used as a source of cheap food. Some countries, such as Japan, already use algae in the form of seaweed in their diets.

Algae are found in fresh water, salt water, and even in damp spots on land. The green coloring on moist tree trunks and rocks is a kind of alga. Many kinds of algae are very small, but some kinds, such as the giant kelp, are as big as trees.

One easily collected kind of alga is called **spirogyra.** This is a stringy green plant that you can find on the surface of fresh-water ponds. Bring some home in a jar with some pond water. Snip off a small piece of the plant to put on a slide. Add a drop of water and a cover slip. Under a microscope you will see a lovely spiral band of green chlorophyll in each cell.

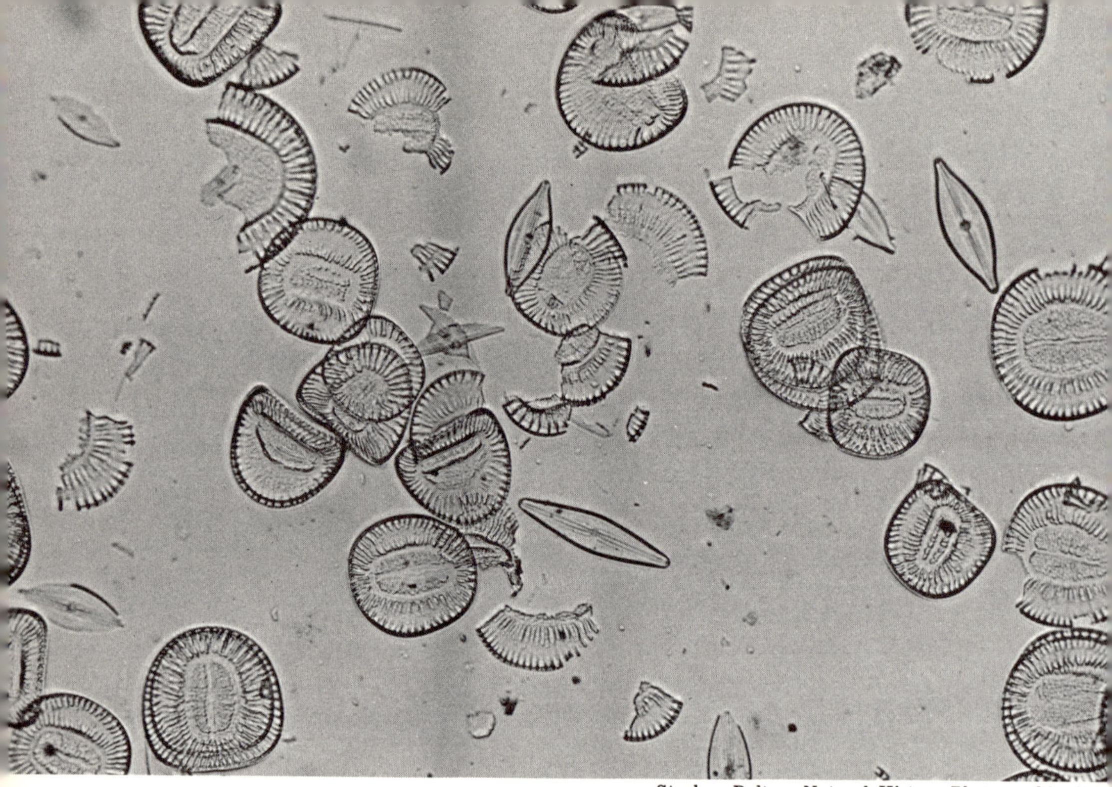

Stephen Dalton, Natural History Photographic Age

Diatoms at 180x.

Certainly the most beautiful of all microscopic algae are the **diatoms.** Diatoms have shells made of silica, the same hard material of which sand is made. After a diatom dies, its shell remains. Diatom shells come in thousands of different shapes and patterns—triangles, stars, circles, ovals.

The shell remains of these tiny plants make up a material called diatomaceous earth. This is sometimes used as the scraping material in toothpaste and silver polish. You can find living diatoms in the brownish covering on rocks or plants at the edge of ponds or streams. Smear some of this material or some silver polish on a slide for viewing.

An onion is a good plant to look at under the microscope. The cells of an onion skin are easy to see and a good object on which to try out something new—staining. Cut a small square through several layers of an onion and remove them. Separate one of the layers and peel off a bit of the skin from the inside. Spread out the skin on a slide, add a drop of water and a cover slip. Under a microscope you will see rows of box-like cells.

Inside each cell you may see a little bubble. This is the **cell nucleus.** You can make it stand out more clearly by **staining** the cells. Add a drop of tincture of iodine (which you can buy in a drugstore) to the edge of the cover slip. The iodine flows under the slip and stains the onion skin. The nucleus becomes yellow, then brown, but the rest of the cell remains light-colored. Now you can see the nucleus easily. Staining helps us look at many things under a microscope that would otherwise be very difficult to see.

Onion-skin cells, stained, at 110x. A second layer of cells shows through faintly from underneath.

Clay-Adams

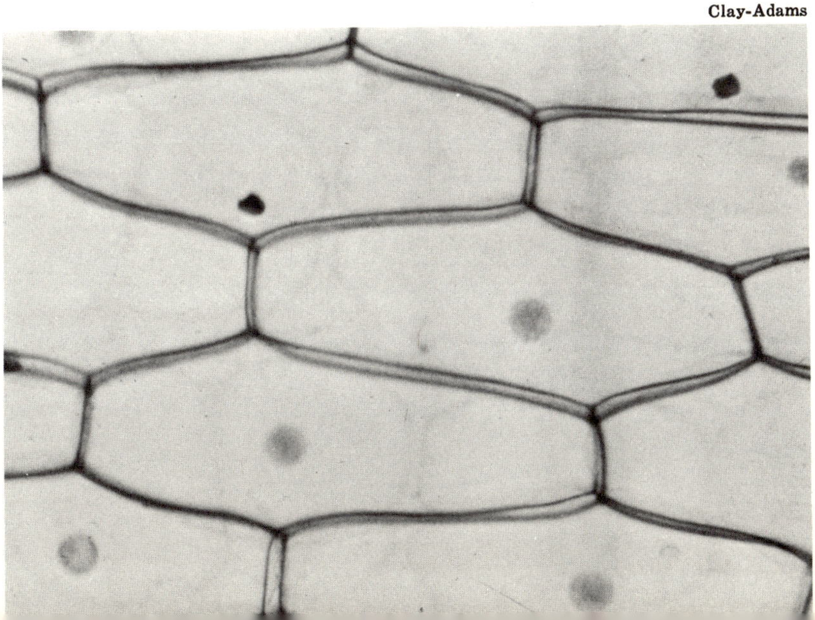

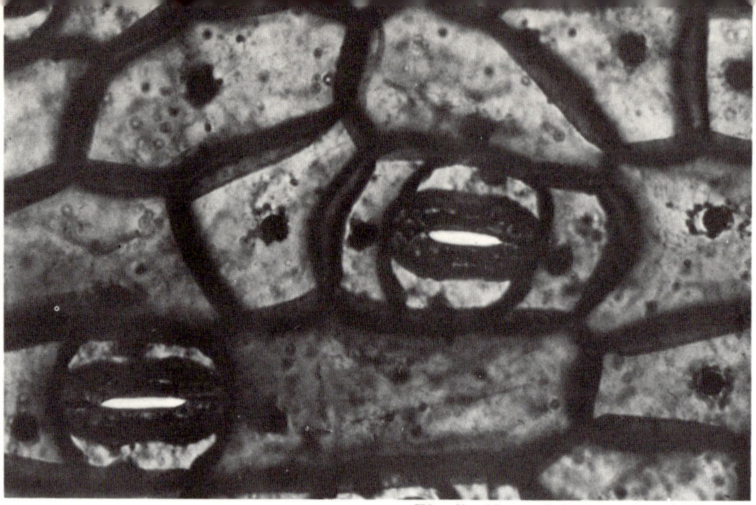

Ward's Natural Science Establishment

Leaf-skin cells from a spiderwort plant, at about 250x. The white openings are stomates.

You can see other kinds of cells in the underside of a leaf. Use a thick leaf from a plant such as a geranium or a coleus. Roll the leaf between your fingers until the skin is loose. Peel off a small piece of the skin on the underside of the leaf. Mount it on a slide with a drop of water and a cover slip.

Look for a pair of bean-shaped cells surrounding a small opening. These cells are called **guard cells.** The opening is called a **stomate.** The plant gives off excess water through the stomates. The guard cells swell or shrink, closing or opening the stomates, depending on the amount of water that must be given off. The water is usually given off as a gas, water vapor. Another gas, carbon dioxide, is taken in through the stomates. Carbon dioxide is used by the plant to make food.

There are many other things to look at in the world of plants: the parts of a flower, tiny seeds, mosses, fern spores, water plants. Keep notes and drawings of what you see. They will help you in the future and will let you share your discoveries with others.

LARGE ANIMALS, SMALL CELLS

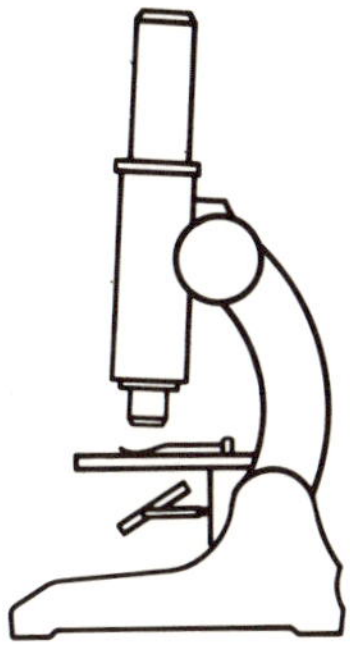

You cannot look at a whole large animal under a microscope, of course. But you can look at the cells of a large animal. In many-celled animals, each part of the body has a special job to do. The cells that make up that body part are often different from cells in other body parts. We can learn much about a body cell and its job by observing it under a microscope.

Making a slide of body cells is not as easy as making a slide of one-celled animals. In order to see many body cells, you have to separate them into thin layers and stain them. Some cells need very skillful handling. Prepared slides of these cells are sold by biological supply houses (see the list in the appendix). But you can make many kinds of slides by yourself.

Let's begin with the cells lining the inside of your

mouth. Scrape off a little material from the inside of your check with a wooden matchstick or the broad end of a toothpick. Place the scrapings on a slide with a drop of water and a cover slip.

Under a microscope you will see flat, irregular-shaped cells with rounded corners. These are skin cells like the ones that cover your body. They are called **epithelial** cells. On your body the cells fit together like the bricks in a building.

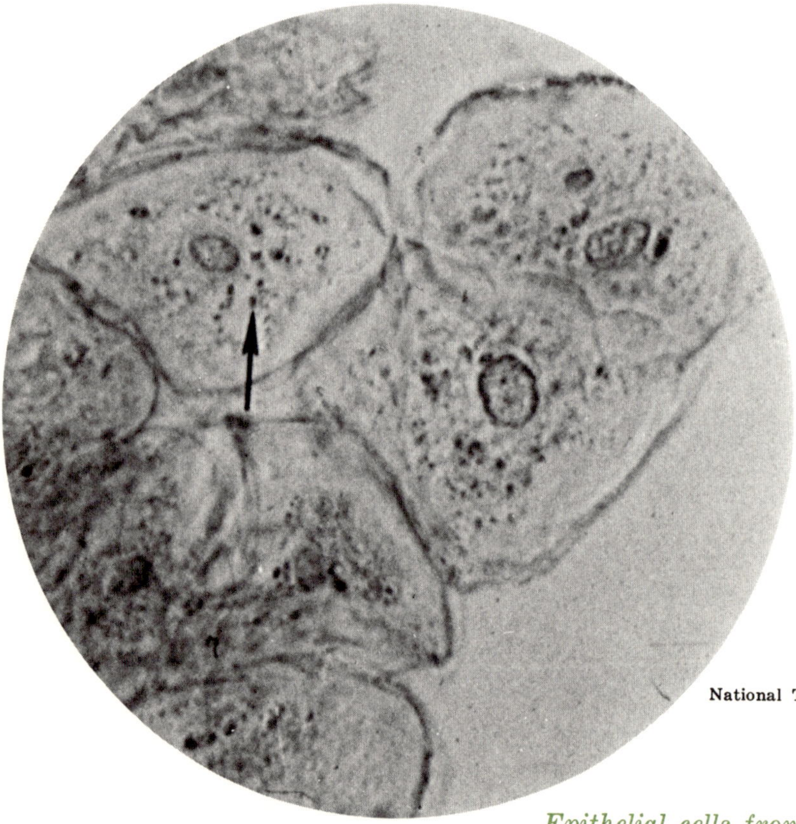

National Teaching Aids

Epithelial cells from the inner surface of a person's cheek, at 900x. The arrow points to dots that are probably bacteria on the surface of the cells.

If you cut down the light coming into the microscope by using a smaller diaphragm opening or adjusting the mirror, you may be able to see the little bubble-like nucleus in each cell. The nucleus can be more easily seen if you add a drop of tincture of iodine to the cells, as we did with the onion cells in the last chapter. This stains the nucleus of each cell brown. Scrapings of loose or peeling skin from your feet or other parts of your body will show epithelial cells very much like those on the inside of your cheek.

Blood cells are more difficult to look at than cheek cells, but well worth the effort. To prepare a blood smear you will need an alcohol lamp, which you can buy from a supply house. **Do not work with fire unless you have your parents' permission to do so.** You can get a drop of blood from yourself or from an obliging friend. Here's how: Wash your hands with soap and water. Take a pin or needle and place the point over the flame of an alcohol lamp until it is almost too hot to hold. Put out the lamp and let the pin cool a moment. Clean a finger with some absorbent cotton dipped in alcohol, then prick the finger quickly. Squeeze the finger until a drop of blood appears.

Place the drop of blood at the end of a very clean slide. With the edge of another slide push the blood into a very thin layer. Spread the blood so thin that there appears to be no layer at all. If it looks reddish, it is too thick. Blood contains so many cells that a thick layer just jumbles them together and makes them difficult to see.

Making a blood smear with two slides; side view.

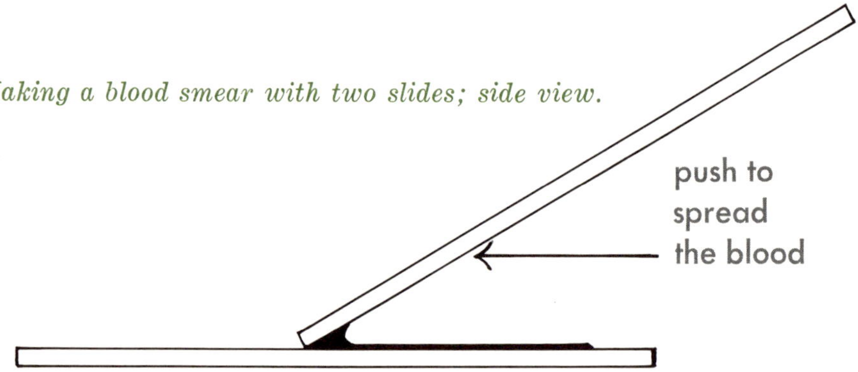

Human blood cells, stained, at about 600x. The large, dark-looking cells are white corpuscles.

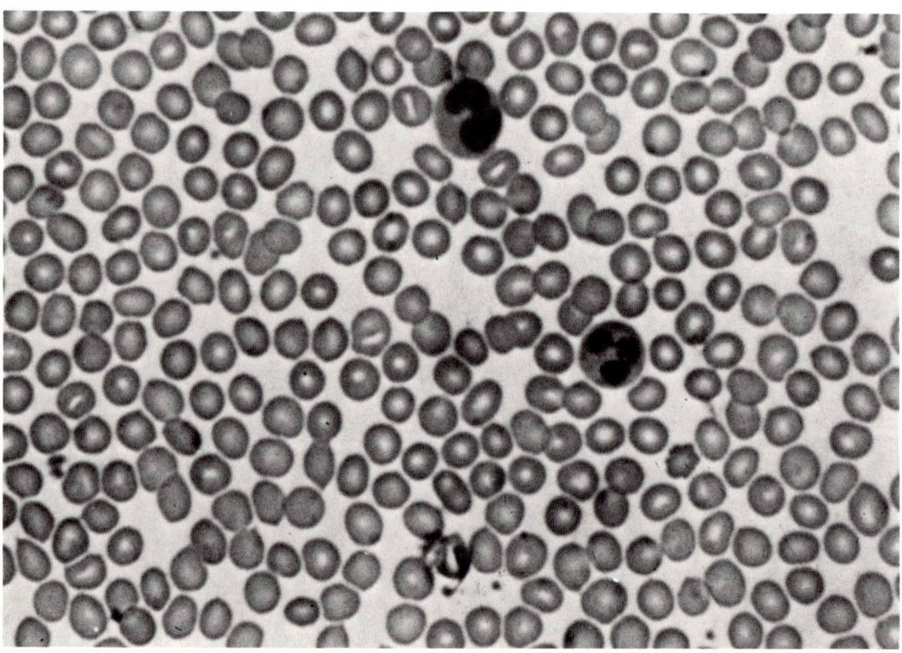

Courtesy Carolina Biological Supply Co.

Let the slide dry and examine it under a microscope. The slide does not need a cover slip. You will see masses of blood cells. Most will look slightly pink or pinkish-yellow. These are the **red blood corpuscles.** They carry oxygen to the other cells of your body. The great number of red corpuscles give blood its color. Unlike most other cells, red corpuscles have no nucleus when mature. The round shadows you see in the red blood cells result from their doughnut-like shape.

You will find a few larger, paler cells in your blood smear. These are **white blood corpuscles.** There is only one white to every 600 reds, so you may have to move the slide to see some. White cells aid your body when fighting germs that cause disease. They have a nucleus but it is difficult to see without staining. To stain the slide, hold it by the edges at one end and carefully pass it once through the flame of an alcohol lamp. This will "fix" the blood smear in place. Then cover the blood smear with permanent blue ink. (Wright's Stain is better if you can buy some in a hobby store or supply house or get it from school.) Let the stain stand for a few minutes and then gently rinse it off. The stained cells will remain on the slide.

You can see blood circulate in the tiny blood vessels in the tail of a goldfish. You will need a small goldfish or other small fish. **If you handle the fish carefully it will not be injured.** Keep the head and body of the fish wrapped in some moist cotton or gauze. Leave only the tail exposed. Place the fish on a flat glass plate or in a petri dish (a low glass dish used in laboratories). Cover the tail with a glass slide and place it under a microscope. Get ready to readjust everything when the goldfish flips its tail!

The tail of a goldfish is almost transparent. You'll see blood cells moving through a network of tiny blood vessels. These tiny blood vessels are called **capillaries.** Larger blood vessels run alongside the tail fin. These are small **veins** and **arteries.** Arteries carry blood from the heart; veins carry blood back to the heart. Blood flows swiftly in arteries. You may see them pulse. Blood flows more slowly and steadily in the veins. Capillaries connect the arteries to the veins.

From time to time drop some water on the fish's tail. Make sure the cotton around its head and body stays moist. After five or ten minutes, return the goldfish to water.

You can see other kinds of cells if you have a small piece of raw meat, such as beef or pork. Remove a bit of white fat from the meat and spread it flat on a slide. Add a drop of water and press the fat down with a cover slip. Fat cells are large, ball-shaped objects.

Use a needle to tease apart a small bit of the red meat. The red meat is muscle tissue. Put the bit of meat on a slide and spread it out. A muscle cell has more than one nucleus. Try an iodine or a permanent red ink stain on the meat to make the many nuclei in each cell stand out. Under a microscope the muscle cells look like long fibers. Many little cross lines, called **striations,** run across the long fibers.

Voluntary muscle cells at 180x. The tiny cross lines on the fibers are the striations.

Stephen Dalton, Natural History Photographic Agency

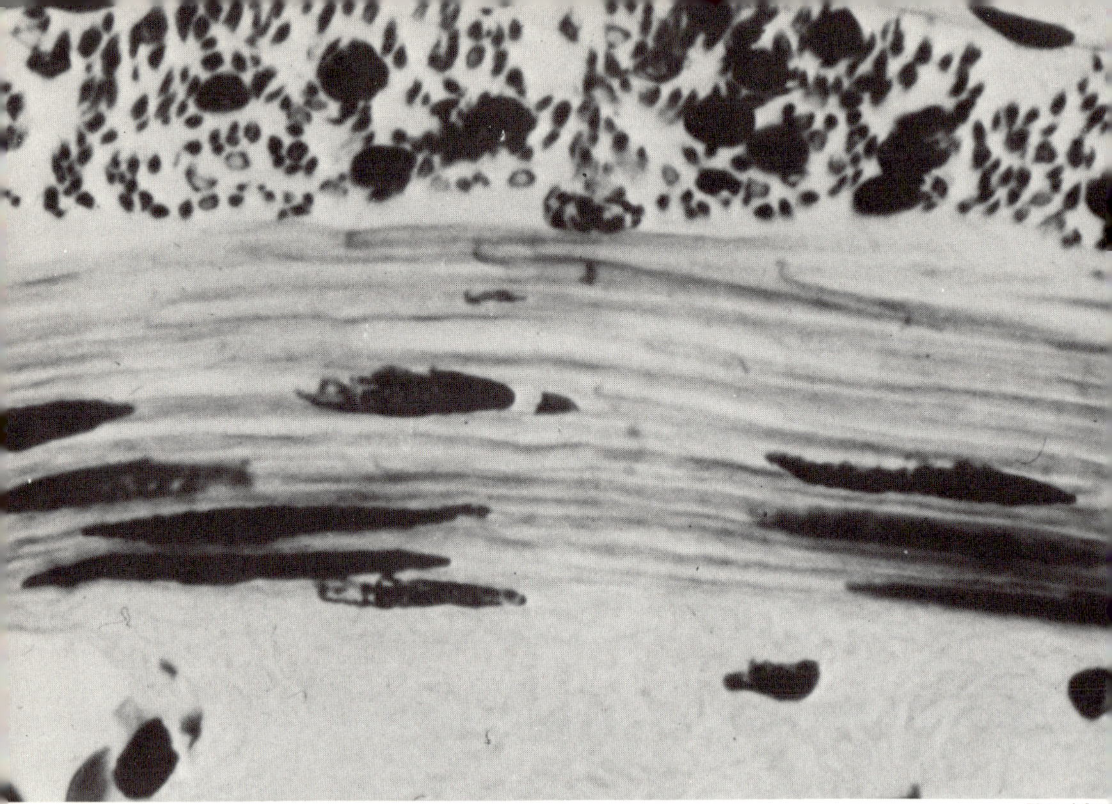

Clay-Adams

A thin section of an intestine at 150x, showing involuntary muscle cells in the center band.

These cross lines are found on all muscle cells that you can directly control, such as your arm or leg muscles. These are called **voluntary muscles**. Smooth muscle cells without the cross lines are called **involuntary muscles.** They are found in such places as the stomach and the intestines. Smooth muscle cells are short, thin, and usually wrapped around other tissues. Perhaps your school science laboratory can give you a piece of the intestines of a fish or frog so that you can see smooth muscle cells under your microscope. Butcher shops may have on hand the intestine and stomach of various animals.

You need special slicing equipment to make a slide showing bone cells, but you can easily make a slide of **cartilage.** Cartilage is the bendable whitish material found in joints, places where two or more bones are connected. You can find some cartilage at the ends of chicken or beef bones. Carefully slice off a very thin section of cartilage with a scalpel or sharp knife. **Always cut away from your fingers when you slice.** Stain the cartilage with some iodine and look at the edge of it under a microscope. You will see irregularly shaped cartilage cells singly and in pairs, surrounded by white connective tissue. The cartilage cells make more of this hard white tissue as they grow.

Cartilage cells, stained, at 100x.

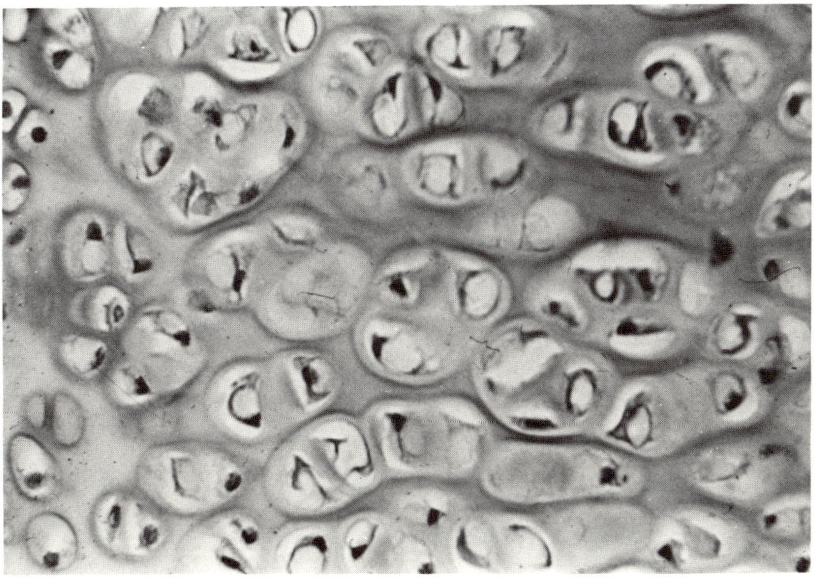

Clay-Adams

Look at any other parts of animals that you can find. Try different stains on each of them. Keep records of which stains work best for each part. Look at the skin of a chicken, bits of its liver or lung, a section of a blood vessel or a piece of its heart. Look at pieces of any dead animals that you may find: a fish, a frog, an earthworm, a snail. Look at scrapings from a turtle shell, an old snakeskin, parts of insects. You'll learn how to make permanent slides of some of these things in the next chapter.

LOOKING AT INSECTS

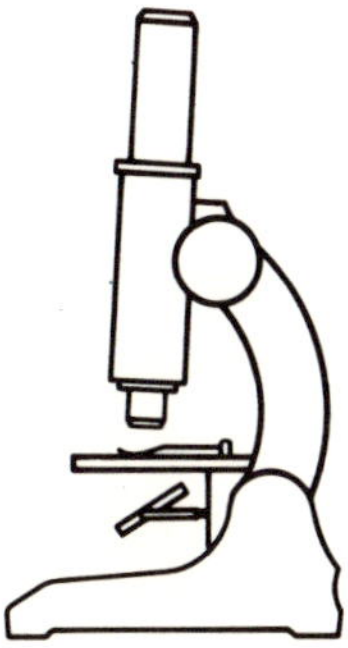

Ants, bees, wasps, butterflies, moths, beetles, fleas—insects are all around us. In fact, there are more kinds of insects than all other kinds of animals put together. Insects are easy to find and interesting to look at with a microscope.

Let's begin with an insect that you can catch indoors as well as out-of-doors: the fly. Catch one in your hand or trap one in a jar. You will need a method of killing flies and other insects as quickly and painlessly as possible. Use a small jar with a tight-fitting cover. Place a plug of cotton at the bottom of the jar and add about 10 drops of household ammonia or a cleaning fluid such as Carbona. **Do not inhale the fumes in a confined space, such as a room with closed windows.** Replace the top of the ammonia or Carbona bottle immediately. Put the fly in the jar and cover it tightly. Do not remove the fly until it is dead.

Stephen Dalton,
Natural History Photographic Agency

A fly's leg at 180x, showing the foot pad with sticky hairs at top.

A fly is too large to look at all at once under a microscope, so you will have to look at separate parts of it. Suppose you begin with the fly's leg. Detach the leg close to the body with sharp scissors and put it on a slide. A cover slip is not needed. Use the low-power lens to look at the leg.

You will see that the leg is hairy and is divided into five main parts. At the end of the leg is a foot with some small claws. In the midst of the claws is a soft pad with a number of sticky hairs. These permit the fly to walk on the ceiling and most other surfaces.

In the future you might like to compare the fly's leg with the legs of other insects. The best way to keep the leg is by mounting it permanently on a slide. A collection of permanent slides is a good record of what you have seen under a microscope.

There are several different ways to make permanent slides. The method you use depends on the object you are mounting and how long you want the slide to last. One of the easiest methods uses Karo light corn syrup or any clear syrup to mount the object.

Use a very clean slide and cover slip. Place a drop of clear syrup in the middle of a slide. With a pair of tweezers, center the fly's leg in the drop. Stand a cover slip on edge on the slide and gently lower it over the drop of syrup. It should come down like a hinged trap door. If you do it slowly, you will prevent air bubbles from being trapped inside. If any bubbles do form, press them out very carefully to the edge of the cover slip. Any excess syrup can be removed with a damp tissue. Glue a paper label at one end of the slide. Note the name of the object, the date, and the mounting fluid used. Leave the slide in a flat position until the syrup hardens. The Karo slide will last for many months, until the syrup becomes discolored.

More long-lasting mounts can be made with Canada balsam or other mounting mediums sold by a biological supply house. These fluids provide better results than clear syrup. They dry more quickly. They remain clear and do not change color. Slides made with them will last for many years. To clean excess balsam from the slide (and your fingers) you will need a solvent called xylol. **Caution: keep away from fire when working with xylol.** Follow the same directions that you used for making Karo slides but do not clean the excess balsam from the slide until several days later.

Whole, small insects and parts of larger insects may need protection from being crushed by the cover slip. You can use a cavity or well slide to mount them. Or make your own slide. Here's how: In a hardware store, buy a number of different-sized plastic or metal rings or washers. Select a ring of about the same thickness as the insect and glue it to a clean slide. You can use household cement for this job. Place the insect in the hole in the washer, add enough balsam to fill the hole, then put on a cover slip. Round cover slips look neater but are not necessary.

You can make an even better slide if you **clear** the insect before you mount it. Clearing makes it easier for light to pass through and gives a better view. You will need to buy a small bottle of 95% to 100% alcohol from a biological supply house. Mix a small amount of the alcohol with an equal amount of water. Place the insect in the mixture for ten minutes. Then place the insect in a small amount of pure alcohol. After ten minutes, remove the insect and place it in some xylol. You should see the insect looking more transparent after about ten minutes in the xylol. Mount the insect as before. Pour the used alcohol and xylol down the drain.

National Teaching Aids

A grasshopper leg (left) and a diving-beetle leg. They are actually about three-fourths of an inch long.

Dried objects such as fibers and paper can be cleared with xylol without first using alcohol. But because xylol turns milky when mixed with water, alcohol must be used first when clearing all other objects.

If all these steps for making permanent slides seem very complicated to you, make only simple Karo slides at first. If you are interested, later on you can try some of the other procedures.

Here are some other things to look at on your insect slides. An insect has three parts: a **head,** a **thorax** (chest), and an **abdomen** (stomach). Insects have six legs attached to the thorax. Insects are covered by a stiff outside skeleton that supports and protects them.

Examine the legs of different insects. A praying mantis has claw-like legs in front that it uses to capture its prey. A bee's legs have a basket made of stiff curving hairs. The baskets become stuffed full of pollen when a bee visits a flower. An ant has a little comb made of hair on its leg. It uses the "comb" to clean its antennae. Swimming insects, such as the water boatman, have flat legs that serve as paddles. Check also grasshopper and beetle legs.

A honeybee's leg, enlarged here 35 times from its natural size of about one-fourth inch long. The arrow points to the pollen basket.

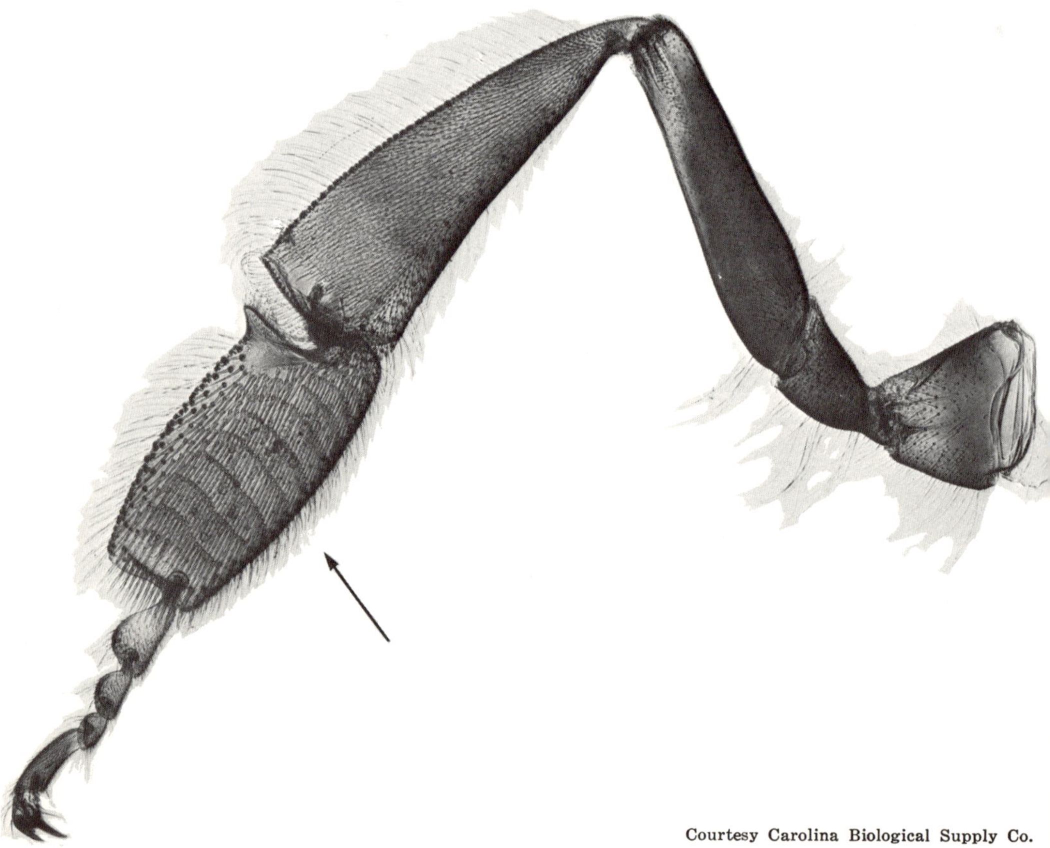

Courtesy Carolina Biological Supply Co.

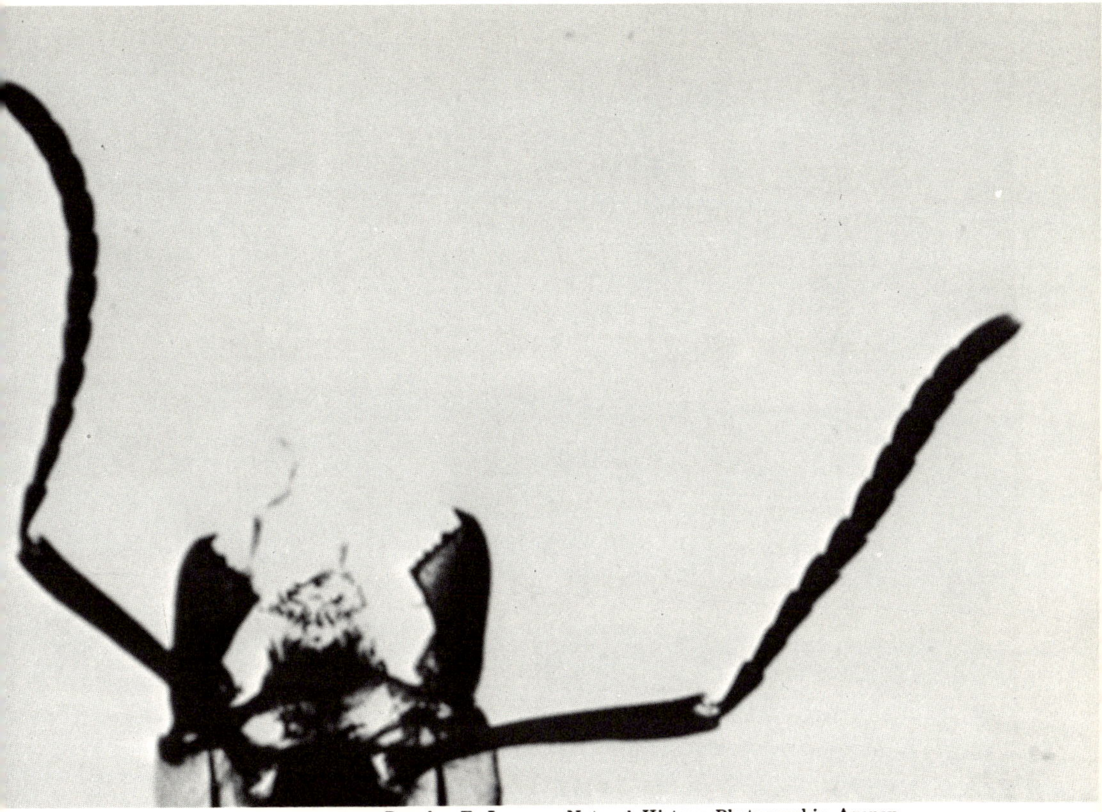

Douglas F. Lawson, Natural History Photographic Agency

The head and antennae of an ant, at 35x. The antennae are actually about one-eighth of an inch long.

The different kinds of **antennae** (feelers) on insects are also good to look at. An antenna is an important sense organ for an insect. It may receive messages of both touch and odor. Watch what happens when one ant meets another ant. They touch antennae to find out by odor whether they are members of the same nest.

There are many different forms of antennae. Some look like little bricks, one on top of another. Some are very feathery. The antenna of a male moth is larger

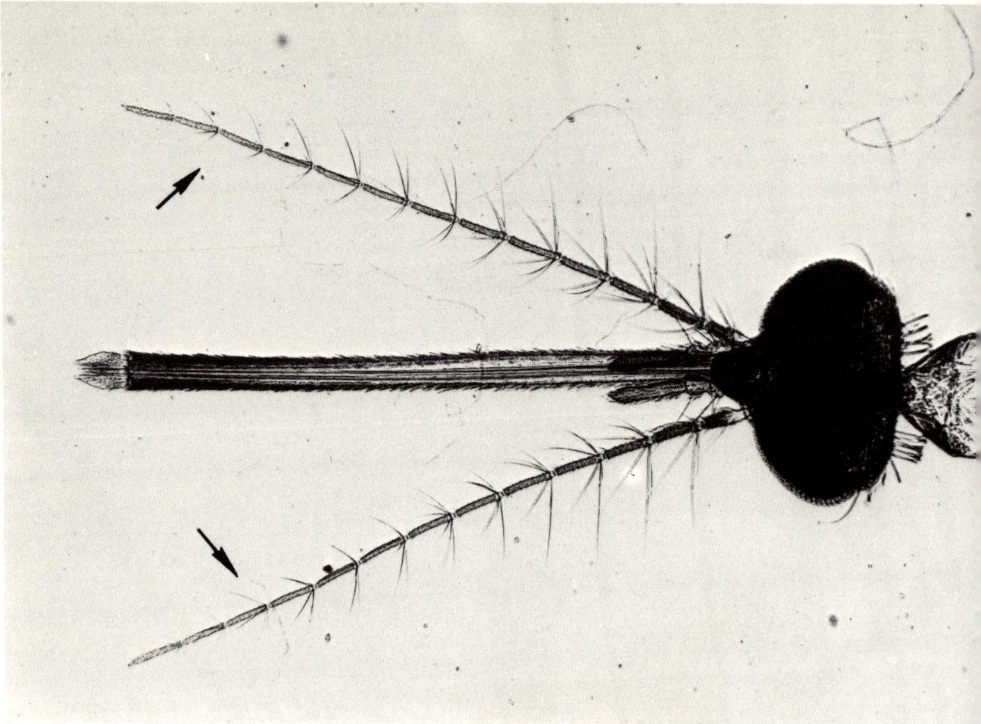

Courtesy Carolina Biological Supply C[illegible]

Mosquito heads at about 60x. Above: female. Below: male. The arrows point to the antennae. In between the antennae are the proboscis (sucking tube) and other mouth parts.

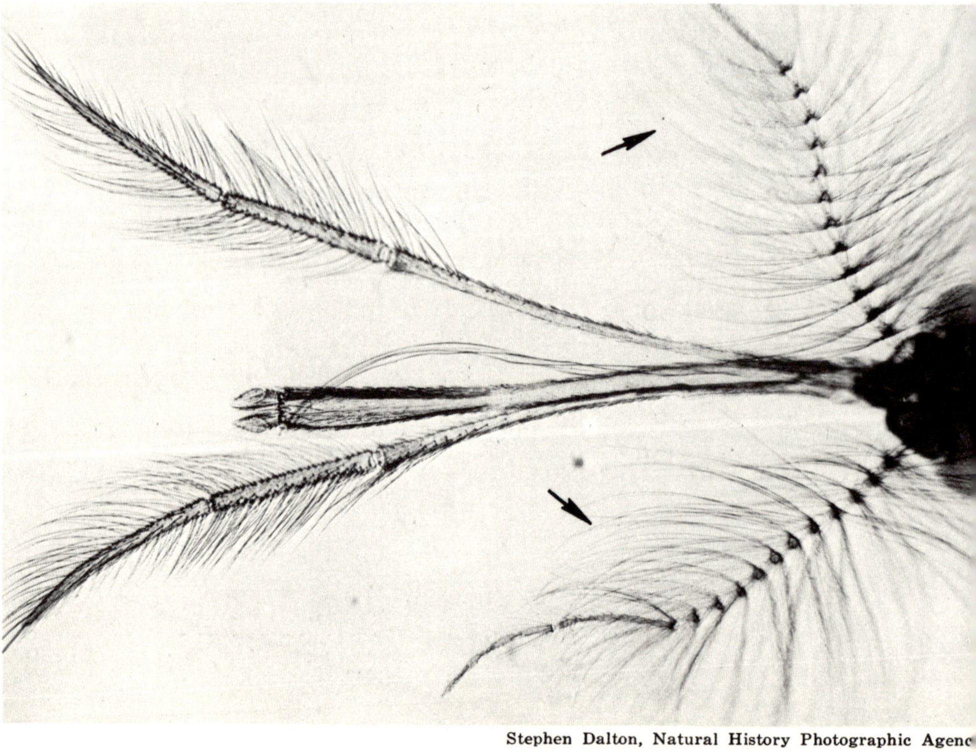

Stephen Dalton, Natural History Photographic Agenc[illegible]

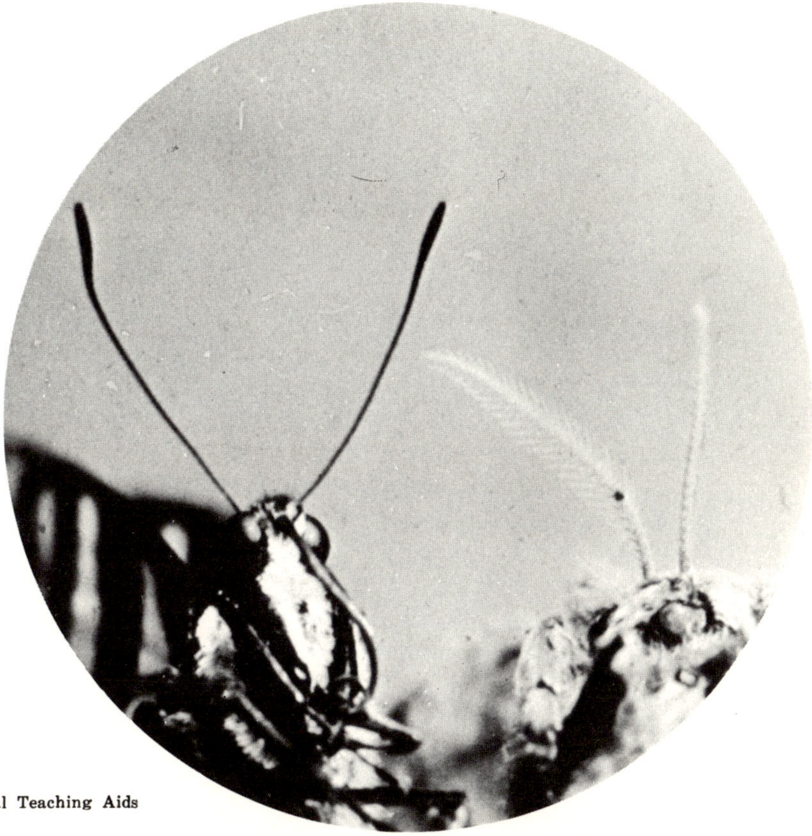

National Teaching Aids

Antennae of a butterfly (left) and a moth.

and more feathery than that of a female moth. The male moth's antenna can detect the odor of a female moth from far away. The male mosquito's antenna is also more feathery than the female's antenna. But it's not odor that the male mosquito's antenna detects, but the sound that the female makes!

You can also compare the eyes, mouths, and wings of

different insects. All of these parts can be detached with a scalpel or scissors. Look at the long, coiled tongue of a butterfly, the stinger of a wasp, the sucking tube of a mosquito. Spiders, eight-legged animals that prey on insects, are also good subjects to look at.

A butterfly tongue at 94x.

Douglas F. Lawso
Natural History Photographic Agen

MORE THINGS TO DO

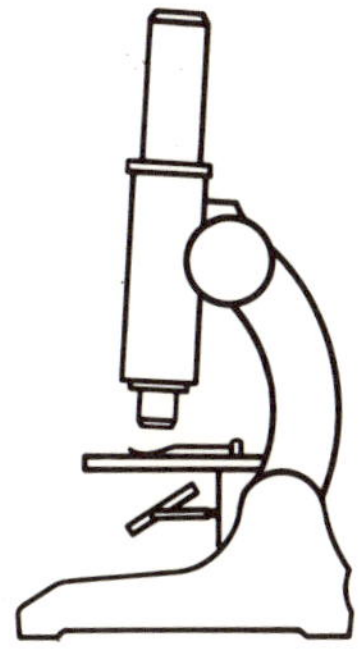

There are so many things to look at with a microscope that it is difficult to decide which ones to suggest. Here is a roundup of some ideas.

Scientists say that no two snowflakes are exactly alike. If that is so, then the ones you examine under a microscope have never been seen before! In any event, snowflakes are beautiful to look at.

During a snowfall, take a microscope and a piece of black cloth or a few glass slides to a protected spot out-of-doors. If you are willing to leave open the window in your bedroom, you can work there. After your equipment has become chilled, catch some snowflakes on the cloth or the glass slides. Look at them under the low-power lens. Don't breathe on the flakes or they will melt.

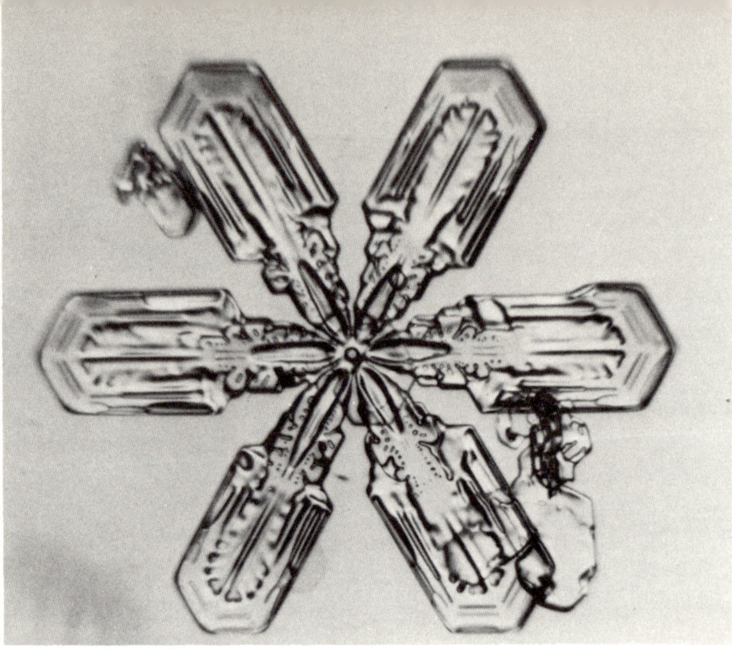

Snowflake crystals at about 25x.

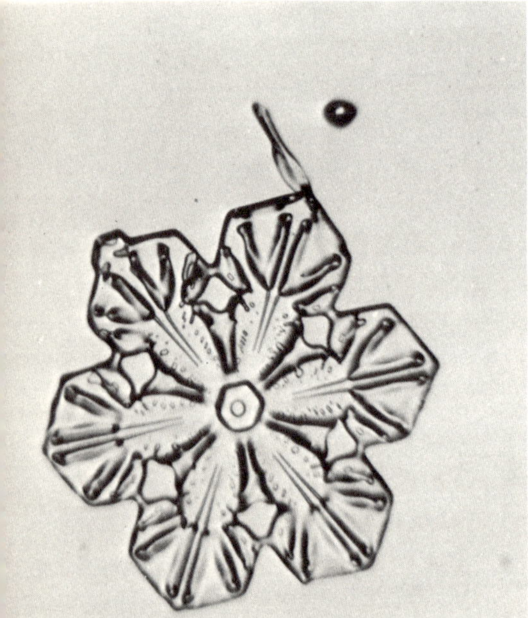

Courtesy Carolina Biological Supply Co.

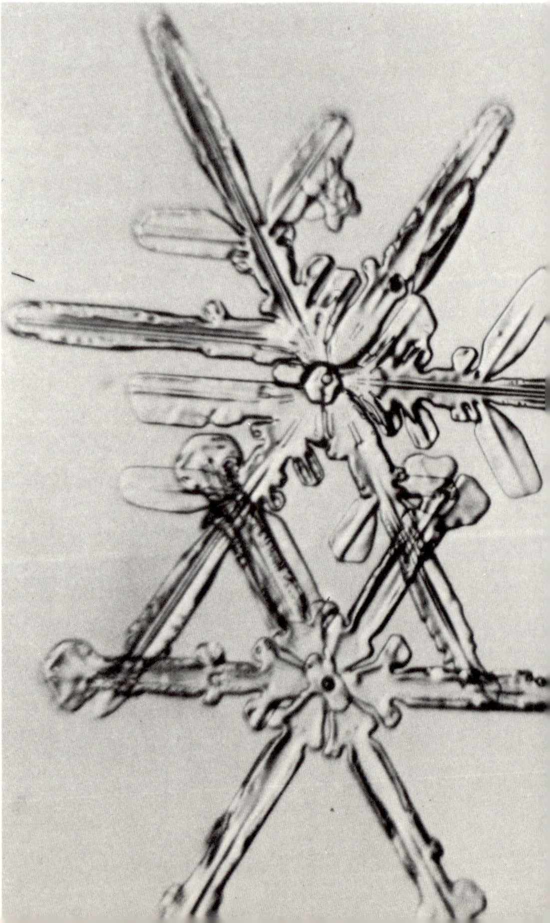

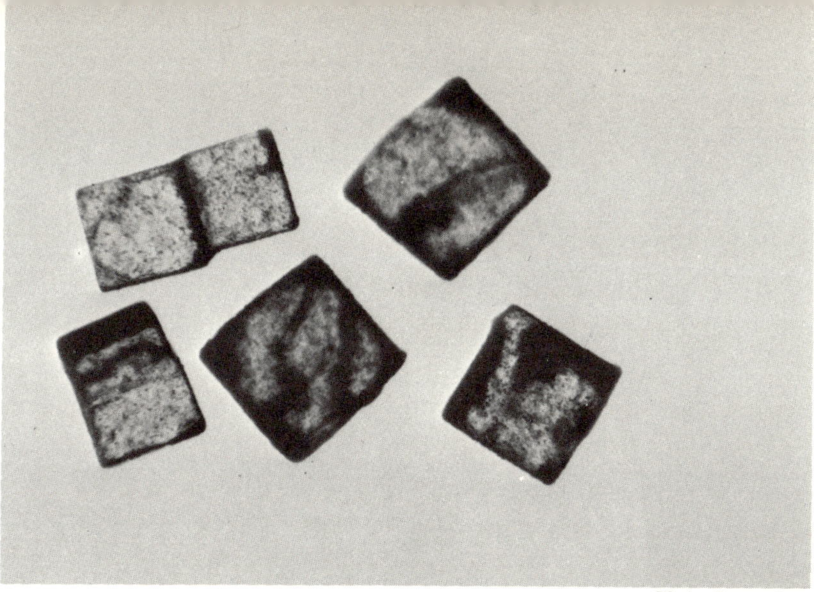

Photo Researchers

Salt crystals at about 50x.

Most small, hard snowflakes are six-sided. Large, soft flakes are partly melted and difficult to make out. Make drawing of the flakes if you can. See if any of the flakes are alike.

Under a microscope you can watch how salt crystals form. Dissolve a teaspoonful of some ordinary table salt in a half cup of water. Place a drop of the solution on a slide and examine it under low power. (Omit the cover slip.) As the water evaporates, you will see crystals of salt forming and growing larger. Cut down the light by moving the mirror and see how the crystals glow against a dark background. Salt crystals are whitish square blocks. For other shapes and colors try chemicals such as potassium permanganate, alum, copper sulfate, or boric acid. You can buy these in a hobby store, drug store, or supply house. Perhaps your school can give you some.

Most people connect **bacteria** with disease. This is only partly true. Some kinds of bacteria do cause diseases but other kinds are helpful. For example, bacteria cause leaves and other materials to decay, improving the soil.

You can find bacteria almost anywhere. Try taking some scrapings from your teeth with a toothpick. Put a small bit of the scrapings on a slide. Add a drop of water and spread it around. Hold the slide with forceps and pass it through the flame of an alcohol lamp several times. (Don't work with fire unless you have your parents' permission.) Let the slide cool. Spread permanent blue ink on the slide and let it stand for a few minutes. Rinse the slide gently and let it dry. Bacteria are very small and you will need high power for viewing. You may need 400X or more to see them clearly.

Under the microscope, you may see three different shapes of bacteria—spheres, spirals, or rods. The spheres are called **cocci**; the rods are called **bacilli**; the spirals are called **spirilla.** Don't be alarmed that you have these in your mouth—everybody does. They will do no harm unless your body's defenses are weakened.

The three shapes of bacteria. Left to right: cocci, bacilli, spirilla. At about 1000x.

Courtesy Carolina Biological Supply

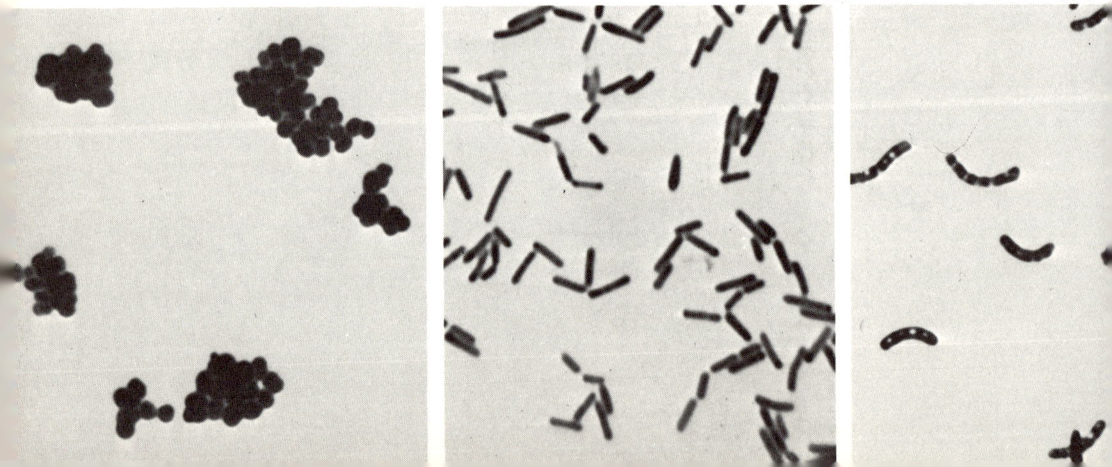

Carl Struewe, Monkmeyer

The interlocking branches of a hummingbird's wing feather at 120x.

Look at a feather under a microscope. The dark, central shaft grows from the skin of the bird. Interlocking branches form a tight, waterproof covering. Feathers have two important uses for a bird: warmth and flying. Wing feathers are long and very light. Down feathers are found over many parts of a bird's body and are usually smaller and softer. Some kinds of feathers seem to change color when the direction of the light changes. Try moving the feather to see if that happens.

Eric V. Gravé, Photo Researchers

A fish scale (from a sole) at 50x, showing barbs and growth rings.

Just as feathers protect a bird's skin, so do scales protect a fish's skin. Scales are made of a colorless, hard material. Some scales are thick in the center and thin at the edges. They overlap each other like shingles on a house. The rings on the scales can sometimes be used by scientists to tell the age of the fish. Little hook-like barbs at the edge of some scales show where the scales were attached to the fish.

Place a few radish or grass seeds under a damp blotter in a jar. In a few days they will sprout. Place a sprouting seed on a slide and examine it under low power. You will see that the fuzz along the roots is really thousands of tiny roots, called root hairs. Water and dissolved minerals pass into the plant through the root hairs.

The number of roots and root hairs on a plant is incredible. Some scientists once grew a single rye plant for several months. At the end of that time, they removed the plant from the soil and carefully measured the length of each of its roots and root hairs. They found that the plant had 7,000 miles of root system!

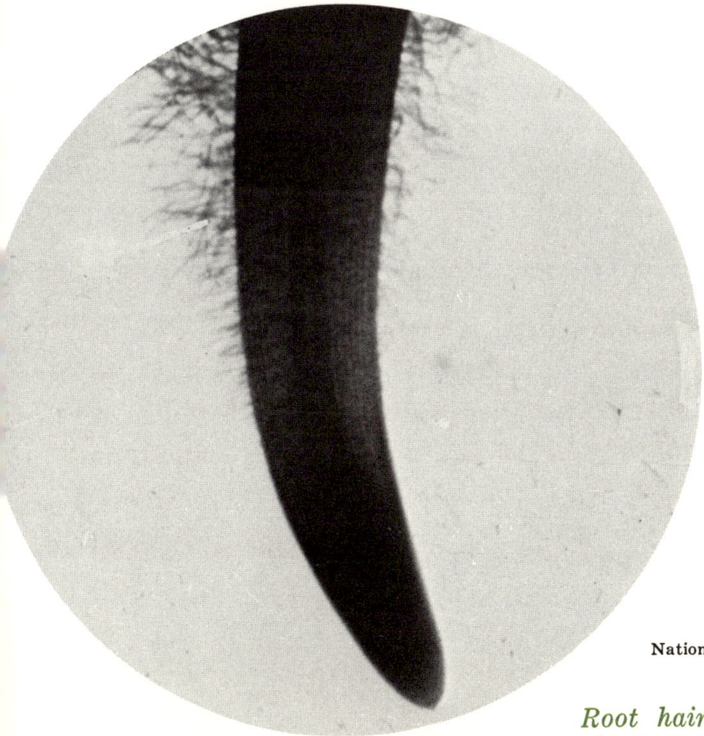

National Teaching Aids

Root hairs on the root tip of a plant, stained, at 20x.

A most remarkable little animal for your microscope is the **tardigrade** or "water bear." You may be able to find one attached to some pond weeds, or in a rain gutter, or even on some damp moss. Since you cannot see a tardigrade with your unaided eye, you will only find it by searching over pond weeds or moss with a microscope. A tardigrade has eight stumpy legs ending in claws. It stumbles over objects in an amusing way.

The tardigrade can live under the most extreme conditions. It remains alive if the temperature falls far below freezing or rises over the boiling point. It can exist without air for days. If it dries out, that's no problem either. Put it into water again even a year later and it will be fine. The tardigrade is one of the toughest creatures around.

The sea is home for great numbers of microscopic living things called **plankton.** Plankton can be collected in a fine nylon net. Cast your net anywhere in the sea, then shake it into a glass jar containing sea water. Make sure that no metal touches the water. Metal salts dissolving in the water may poison the plankton.

At home, place a drop of the sea water on a slide, use a cover slip, and examine under low power. You will probably see many different kinds of Protozoa, tiny lobster-like animals called copepods, single-celled sea plants, tiny jellyfish, and a host of other plants and animals that make up plankton. You will find something new no matter how many times you look at plankton.

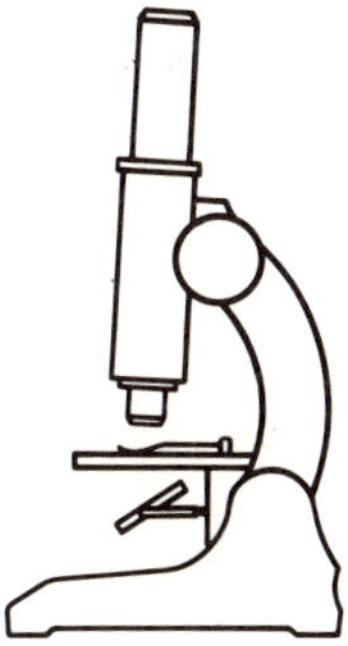

YOUR OWN MICROSCOPE

You may decide that you would like to have a microscope of your own. A microscope is a worthwhile investment. It can provide hours of enjoyment. With proper care it can last a lifetime. If you receive a microscope as a gift, it will probably last far longer than any other gifts you receive.

A decent microscope usually costs at least fifteen or twenty dollars. Most sell for a lot more. Some toy microscopes sell for much less—about five or six dollars. These are usually not very good. Their poorly made lenses give a blurry image at any magnification. You might get so discouraged using such a poor instrument that you would give up microscope fun entirely. Save your money until you can buy a better one.

Try to buy a microsope from a store that specializes

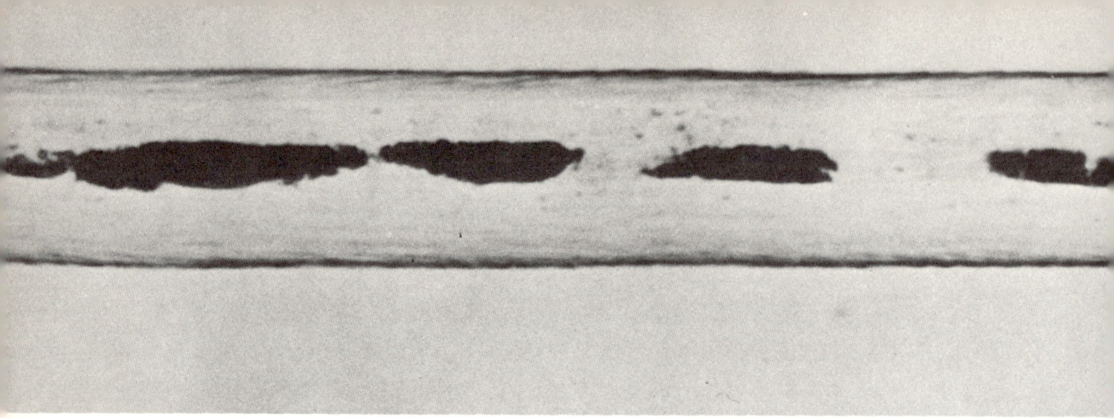

Eric V. Gravé, Photo Research

Carl Struewe, Monkme

More things to look at: animal hairs. Top to bottom: human hair at 260x, mouse hair at 400x, Persian lamb hair at 260x.

Eric V. Gravé, Photo Research

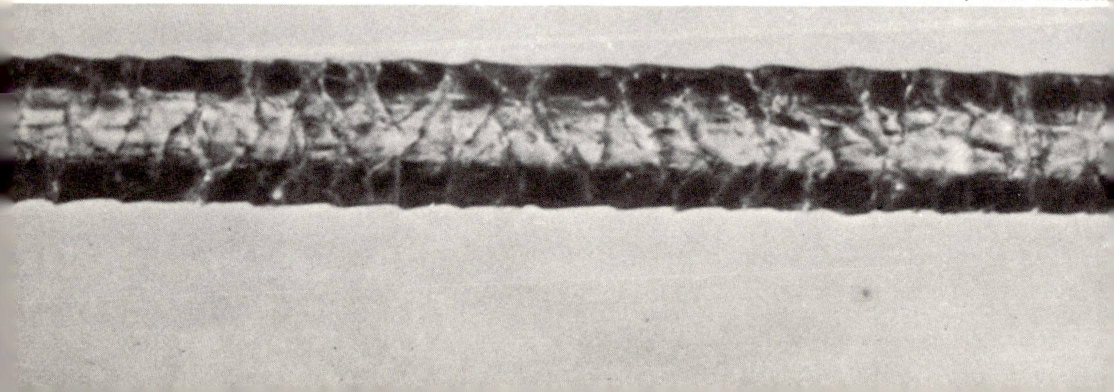

in them—a scientific supply company or a camera company. Otherwise, look into department stores, toy stores, and hobby stores. Try out several different models. Ask if there are any good, used microscopes available. Look at some sample slides through each microscope.

The focusing knobs on the microscope should move smoothly and evenly. When looking at a slide, make sure you see a clear image without color fringes along the edges. Better-made miscroscopes use **achromatic** lenses. These lenses do away with fuzzy color fringes and give a sharper image. Some inexpensive microscopes are made to look attractive with all sorts of fancy gadgets. They may magnify up to 600X. But a less-powerful achromatic microscope is far more useful than a fancy-looking nonachromatic one.

If you cannot visit a store in person, write for a catalog to some of the supply companies listed in the appendix. Many companies have suitable microscopes at different price ranges. Perhaps your science teacher can help you in choosing one. Also buy a supply of glass slides and cover slips. Useful instruments are a pair of tweezers or forceps, a dissecting needle, and a scalpel.

Here are some rules to remember when handling your microscope:

Cover your microscope when not in use to prevent it from getting dusty. Keep it in its case or under a plastic cover.

Wash your hands before you handle your micro-

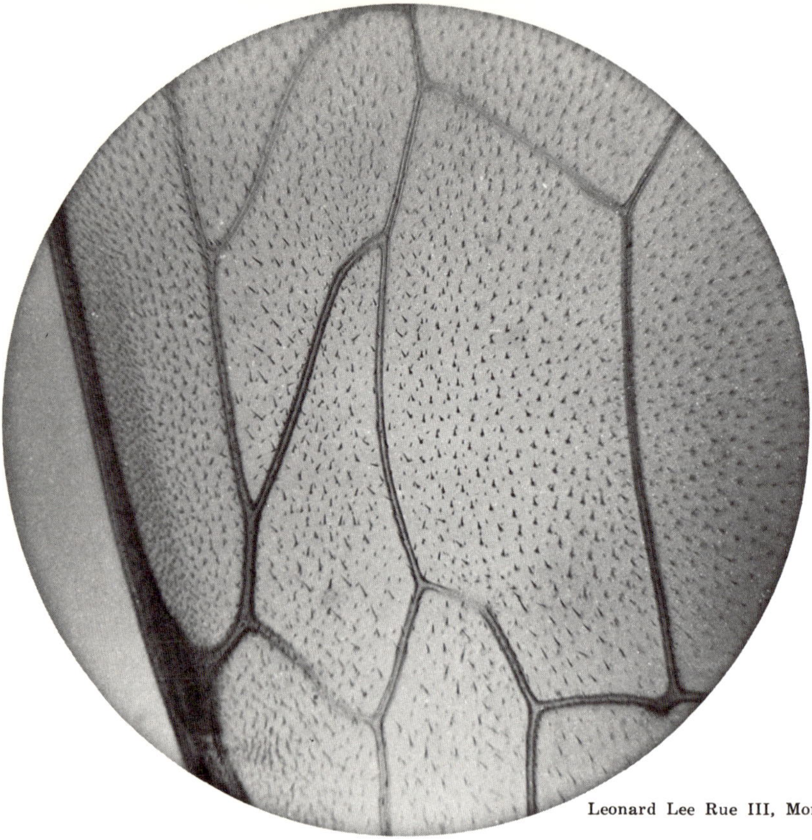

Leonard Lee Rue III, Monkmeyer

More things to look at: insect wings. Above: a fly's wing at about 100x. Below: scales on the wing of an Apollo butterfly, at 200x.

Carl Struewe, Monkmey

scope. Do not touch the lenses with your fingers. This gets oil and dust on them. Keep the lenses clean by gently wiping them with lens paper, which can be bought in a camera store.

Use a firm, steady table when working with your microscope. A wobbly table shakes the image and makes it hard to see.

Learn to keep both eyes open when using a microscope for long periods of time. If you find it difficult to do this, use your hand to cover one eye while viewing. Look with one eye for a while, and then look with the other. Working with one eye held closed is very tiring. Get used to keeping both eyes open.

If you wear glasses, it is best to remove them when using your microscope. You will see a fuller view as your eye gets closer to the evidence. The microscope lenses help you see things clearly even without your glasses.

Never work in direct sunlight.

Always focus the microscope by moving the body tube upward, never downward.

Be especially careful when working with sharp tools or fire.

After you follow these rules for a while, they will become second nature to you. Your microscope will always be ready to help you explore the fantastic world of the very small.

APPENDIX

MICROSCOPE MANUFACTURERS AND SUPPLY HOUSES

The following is only a partial list of the many companies that sell relatively inexpensive student microscopes and microscope accessories. When writing to any of these places, explain exactly what you are looking for.

Bausch & Lomb, Inc.
635 Saint Paul Street
Rochester, New York 14602

D. P. Bushnell & Company, Inc.
2828 East Foothill Boulevard
Pasadena, California 91107

Carolina Biological Supply Company
Burlington, North Carolina 27215

Edmund Scientific Company
Barrington, New Jersey 08007

Educational Materials and Equipment Company
Box 63
Bronxville, New York 10708

General Biological, Inc.
8200 South Hoyne Avenue
Chicago, Illinois 60620

Testa Manufacturing Company
10126 East Rush Street
South El Monte, California 91733

Ward's Natural Science Establishment, Inc.
Post Office Box 1712
Rochester, New York 14603

Ward's of California
Post Office Box 1749
Monterey, California 93940

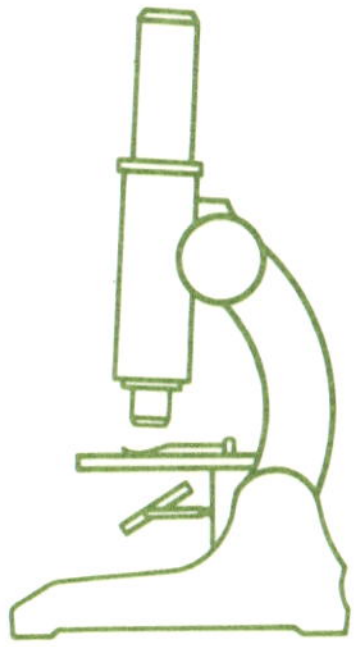

BOOKS FOR READING AND RESEARCH

Anderson, M. D. *Through the Microscope.* Natural History Press, 1965.

Glemser, Bernard. *All About Biology.* Random House, 1964.

Hausman, Leon A. *Beginner's Guide to Freshwater Life.* Putnam, 1950.

Headstrom, Richard B. *Adventures with Freshwater Animals.* Lippincott, 1964.

Jacker, Corinne. *Window on the Unknown: A History of the Microscope.* Scribner's, 1966.

Lindemann, Edward. *Water Animals for Your Microscope.* Macmillan, 1967.

Pinney, Roy. *Collecting and Photographing Your Microzoo.* World, 1965.

Schneider, Leo. *Microbes in Your Life.* Harcourt, Brace & World, 1966.

Selsam, Millicent E. *Microbes at Work.* Morrow, 1953.

Slovan, T. *Under the Microscope.* Childrens Press, 1968.

Zim, Herbert S. and Cottam, Clarence A. *Insects.* Golden Press, 1951.

INDEX

ABOUT THE AUTHOR

Seymour Simon is a science teacher and writer. He has a number of children's books to his credit, including *Weather and Climate* and *The Look-It-Up Book of the Earth.* He has also acted as science consultant and reviewer for various scholastic and children's magazines and has written articles on a wide range of scientific subjects.

A native New Yorker, Mr. Simon graduated from the Bronx High School of Science and City College of New York. He now lives with his wife, Joyce, and sons, Robert and Michael, in Great Neck, Long Island.

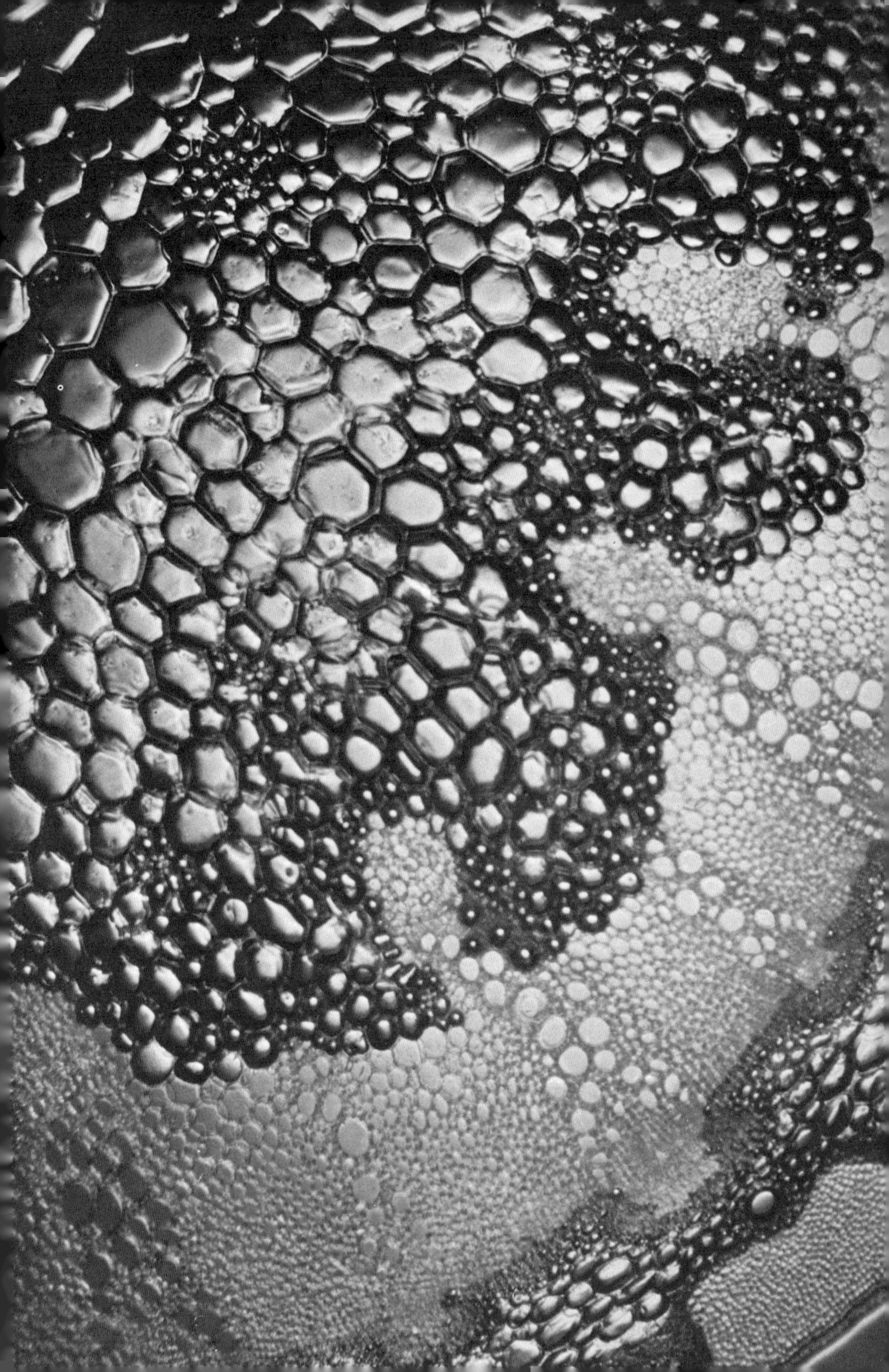